T0220240

EARTH
ENGINEERING

Perspectives, Principles, and Practices

MURRAY SARAFINCHIN

iUniverse, Inc.
Bloomington

iUniverse books may be ordered through booksellers or by contacting:

iUniverse
1663 Liberty Drive
Bloomington, IN 47403
www.iuniverse.com
1-800-Authors (1-800-288-4677)

ISBN: 978-1-4502-7597-2 (pbk)
ISBN: 978-1-4502-7598-9 (cloth)
ISBN: 978-1-4502-7599-6 (ebk)

Library of Congress Control Number: 2010917405

Printed in the United States of America

iUniverse rev. date: 7/7/2011

Preface

Today we are fortunate to have university textbooks and library references to benchmark various aspects of geoengineering. These secure a wealth of detailed information on engineering geology, geotechnical and geoenvironmental engineering, hydrology, hydrogeology, mining, construction quality assurance, site remediation practices, and related topics. Despite this wealth of information, we habitually encounter earthworks and geoenvironmental problems during engineering, construction and mining projects. Problems may develop in the field from incorrect decisions made during the feasibility, planning and technical design stages concerning project specifications and plans, or from schedule and budget constraints during the construction stages. These problems are commonly of a basic, fundamental nature rather than due to more esoteric causes. This book addresses some of the day to day matters encountered in geoengineering. As such, this book is not about theories and formulae in geoengineering. This book uniquely offers a geoengineering collection with stimulating overviews and enlightening insights to our dynamic Earth transformations. It focuses our thinking to a perspectives, principles and practices approach to the investigation, design and construction challenges commonly encountered in earth engineering and environmental sciences.

These chapters transcend several basic and specialty subjects. Earth formation including soil, rock and groundwater components plus classification methods is presented. Various earth structures and ground subsidence are described. Some specialties are specific geotechnical problems such as expansive clays, collapsible soils, and freezing effects that cause distinctive problems in design, construction and service life. Included are helpful chapters on contamination and remediation, geotextiles, landfills, global warming, health, law and ethics.

The objective of this geoengineering sciences reference book is to provide perspectives, principles and practices to understand common geoengineering problems in civil engineering, construction and mining. The intended audience includes geologists, engineers, architects, contractors, developers, regulators, planners, earth scientists, educators and students who must plan, design, build and operate projects with geomaterials, provide drawings and specifications for the work, and then provide onsite engineering reviews that must optimize earthworks. A theoretical treatment of the issues is deliberately avoided, so prior education in mathematics, sciences and engineering is helpful, but not required to make use of this technical reference book.

Due to the uncertainties of earth evolution, including non homogeneous and anisotropic subsurface conditions as related to engineering, construction, mining and environmental sites, it is always best to set a conservative geoengineering course respecting any uncertainties affecting loss of equipment, property damages and most importantly human safety. Consequently, this reference book will be useful to scientists, engineers, contractors and quality control/ quality assurance personnel working with geomaterials. Most projects today require integrated multidisciplinary teams, and it is felt this reference book will prove useful to specialist educators, planners, designers, constructors and practitioners who must work with applied geoengineering and environmental sciences as part of their tools. The student studying geology, civil engineering, architecture, construction, mining and environmental subjects will find this book useful as a guide to the geoscientific background, actual field problems and their solutions, and as a supplement to more theoretical academic studies.

Geoengineering technology is implemented daily to further mankind in the quest for knowledge and sustainable infrastructure development. It is professionally gratifying to be a part of this progressive industry. Hopefully, the perspectives, principles and practices provided in this reference book will assist novices as well as the experienced to build even more upon past experience and to reduce the errors that bring us to where we are.

Whether being more comfortable onsite solving problems for clients or in the office writing reports about them, or sparing time in university libraries the compilation of this book has required an extraordinary effort over many years. Discipline to complete the work has prevailed partly because of my dislike for the confusion created by special interests, a passion to have geoengineering showcased for the prime profession it is, and the fact that it is an essential and rewarding first step for sustainable development in all civil engineering, construction and mining sites.

The impetus to create this first book publication comes from within, and it is enhanced by valuable professional experiences. The uniqueness of such broad based subjects comes from the several and wide variety of Sarafinchin projects, and earlier assignments with Trow, Golder, and Monenco. The compilation of writings commenced with my role as a thirty year speaker for geoengineering seminars to the Ministry of the Environment and the Good Roads Association, plus my special lectures to universities, colleges and conferences. My explorations to Waterloo University, University of Western Ontario, Queens University, McGill University, University of Toronto, Columbia University in New York, Massachusetts Institute of Technology in Boston, University of Berkeley and Stanford University in California, Imperial College in London UK, Florida Atlantic University in Boca Raton, and natural science museums in the world gleaned numerous concepts, beliefs and achievements for this book. I am thankful for these opportunities. Similarly, I appreciate the assistance of my editors and publisher. I am indebted to industry leaders who inspired my early manuscript and later contributed helpful comments. My clients, fellow practitioners, educators, family and friends have been supportive. Time is precious in a consulting engineering office, and although library research and manuscript schedules fell behind, adding new chapter topics begot enlightenment and an amazing stimuli to get the book up and going. There is an addictive thrill in writing it. A variety of colleagues and peers graciously agreed to review and comment on specific portions of the manuscript, and I am grateful for their input.

Members of my staff, particularly my dedicated administrative assistant, are foundations to the preparations and illustrations. Their assistance is greatly appreciated. Any errors or omissions in the text are solely the author's responsibility. A second edition will revitalize our thoughts.

Contents

Chapter I

INTRODUCTION

A ll civil engineering and mining projects engage the surface of the Earth. The feasibility studies, planning, design, construction and operation of such projects are supported by or situated in the Earth's crust.

We are fascinated with, but carefully reminded that every site used for civil engineering and mining structures is unique and nestled within a widely varying Earth terrain. There are often vast subsurface differences, or nonhomogeneous and anisotropic conditions, between adjoining sites due to the underlying and evolving geologic anomalies and the varying stratigraphy. Although site characteristics can vary, the fundamental geoengineering principles applied to the factual observations of the subsurface conditions do not vary.

The renowned engineering geologist, Dr. Robert F. Legget, advises us,

> " ... the records of the geological survey show conclusively that closer cooperation between the geologist and the engineer would be greatly to the advantage of both, and it is a pity that there is no very direct way in which geologists could be kept informed of the progress of important excavations."

Within civil engineering the discipline of geotechnical engineering developed from the science of soil mechanics in the early part of the twentieth century.

The father of soil mechanics, Dr. Karl Terzaghi, stated,

1

> *"...on account of the fact that there is no glory attached to the foundations and that the sources of success or failure are hidden deep in the ground, building foundations have always been treated as stepchildren and their acts of revenge for lack of attention can be very embarrassing."*

and he has gone on to emphasize

> *"...in earthwork engineering, success depends primarily on a clear perception of the uncertainties involved in the fundamental assumptions and on intelligently planned and consciously executed observations during construction. If the observations show that the real conditions are very different from what they were believed to be, the design must be changed before it is too late. These are the essentials of soil mechanics in engineering practice."*

Both the science and the art of geoengineering are changing as many important advances are being made.

The purpose of the first edition of this engineering sciences reference book is to provide an overview of earth engineering and geoenvironmental sciences. It is organized by individual informative chapters with perspectives, concepts, principles, practices, examples, references and suggested further reading. The subjects are explained in simple terms and with nominal theory, formulae and mathematics. This reference book is intended for planners, developers, geologists, geoscientists, engineers, architects, contractors, regulatory agencies, investigators, practitioners, educators, students and interested parties. It embraces the wider specialist field of geoengineering as a marriage of geology and earth sciences with geotechnical and geoenvironmental engineering for civil works and mining applications.

This geoengineering reference book is introductory and an important checklist to the basic areas of interest, the terminology, the concepts, the investigative approaches and practical applications of geoengineering in civil infrastructure and mining projects respecting safe and clean environments for sustainable development. At the beginning of human life on Earth, there were few people, and nature seemed vast and endlessly self renewing. In 1900 almost two billion people on the planet lived mostly in rural village communities. We were agrarians, or farmers, who knew we must depend on and work with Earth's natural processes. Only one hundred years later, in 2000, the global population has grown to six billion, and the number of cities with over one million people had increased to more than three hundred. Most people in industrialized countries live in large cities where it has become convenient to believe that our highest priority is the economy. At the same time, it is important to recognize that every single thing we consume comes from the Earth using geoengineering sciences and goes back to it as waste requiring geoenvironmental engineering solutions. Engineers and scientists in cooperation with owners and industries, are charged with the opportunities to create sustainable Earth developments using knowledge, experience, ingenuity, communication skills and nobler motives.

Chapter 2
OUR EARTH

Our Universe

Our universe or world is the whole of all existing matter, energy and space. From the beginning our ancient ancestors looked up into the night sky, and wondered, where did it all start? Religion has sought to provide us with some of these answers. At the same time a transition to science has developed rational thought to influence our understanding.

Religious Beliefs

The intent of this discussion is not to contradict the religious belief that the universe and the Earth within it were created by God, but to speculate and theorize, based on scientific findings and research by others, just how He made it happen.

In the Bible, Genesis 1:1-5 states that in the beginning God created the Heaven and the Earth. His world is in seven units, or seven days, as follows:

First day, creation of light
Second day, creation of heavens and water
Third day, creation of land and vegetation
Fourth day, creation of bodies of light
Fifth day, creation of creatures of heaven and waters
Sixth day, creation of life on land, vegetable food, and mankind
Seventh day, God rests from all work

Christian tradition has it that Moses wrote the creation story of our Earth along with the first five books of the Bible. The first five books of the Old Testament also comprise the Torah, the sacred scroll that is found in the Tabernacle of Jewish temple. Christianity considers Jesus as the son of God speaking as the Divine on Earth. However, from both the viewpoints of Judiasm and Islam, he is considered a prophet. In Judiasm, there is only one God and the prophets play a key role, and in Islam, there is also only one God and Mohammed is his prophet. Buddhism encourages that all problems on Earth be solved by overcoming greed, hatred and delusion. Hinduism believes in a reincarnation to Earth. Many other spiritual disciplines exist.

All cosmologies, or at least all early conceptions of the universe have a religious aspect. Many used symbolism, plus numerology, cosmology, and astrology thinkers such as Pythagoras, Plato, Ptolemy, Copernicus, and Galileo, while others relied on purely anthropomorphic tales of animal and family interactions

Scientific Reasoning

Cosmology is the study of the evolution and structure of the universe. Science estimates the age of our universe to be between 12 and 15 billion years old.

Cosmogony is the study of the origin of the universe. It probes if we are the only living creatures in this vast cosmos, or are there other beings out there?

In our universe there are many galaxies held together by gravitational attraction. Undoubtedly some have life supporting water and environments which are believed to have extraterrestrial beings different from Earth. If aliens secretly visit Earth, for which there is often reported evidence, then why do they appear to bypass us? Dr. Steven Hawking, the famous physicist, recommends that we do not try to communicate with aliens. Is this because these technologically super advanced beings have no special interest in us, in the same way that we walk through our forests, and neither stop to explore nor engage to interfere with the ants beneath our feet? Are advanced space travelers missing out on the human race in the same way we humans neglect ants, not knowing that genetic and molecular biology studies have shown that ants live in complex societies, they have a caste system, they go to war, they take slaves and they tend gardens on Earth.

Recognizing the 15 billion year age of the universe, the 4.6 billion year old Earth, the estimated 1.5 billion year presence of mankind, the scientific advancements that humans have made in only the past 100 years, and how much more there is to learn in science and engineering, perhaps our best current role is to continue collecting geoscientific knowledge and geoengineering experience for an effective future rendezvous in space?

Scottish geologist James Hutton (1790), one of the founders of earth science stated that geological investigation of Earth showed no vestige of a beginning, and no sign of an end. One of the great contributions to earth sciences is by Charles Lyell in his Principles of Geology (1830) suggesting that our Earth is shaped by constantly acting natural processes with a place for life in nature.

Many geoscientists accept some version of the big bang theory. At first all energy and matter was closely concentrated. About 15 billion years ago a vast explosion scattered everything through space. The Sun is believed to have come from an exploding supernova in our Milky Way Galaxy. Cosmologists believe the universe was over 10 billion years old when our solar system formed 4.6 billion years ago. The gravitational pull of the Sun caused elemental debris, or planetesimals, to orbit around it. Earth started as one of these planetesimals gaining mass and size as other planetesimals collided into it as it orbited the Sun. Each time an asteroid sized planetesimal collided with Earth, the tremendous impact created heat so intense that Earth's surface stayed in a constant molten state. The entire Earth was covered by a sea of molten lava. Some researchers estimate the magma ocean to have been thousands of kilometres deep. The process of planetesimals crashing into each other to form planets is called accretion. Most of the planetesimals that orbited the Sun have accreted into planets and their moons. Consequently we are not in much danger of an asteroid hitting Earth and causing mass destruction. Planetesimals still fall to Earth, however most of them burn up in Earth's atmosphere before reaching the ground. Seen in the night sky these shooting stars are meteors, and if they penetrate through Earth's atmosphere hitting ground, then they are meteorites.

Thus our solar system comprises the Sun and the eight planets plus dwarf planets formed from a cloud of dust and gas spinning in space. The largest and most influential body is the Sun, a glowing ball of gases a million times the volume of Earth. The Earth's radioactivity caused the surface to melt. Upon cooling molten rock layers formed the continents and volcanic gases formed the atmosphere. Water vapour condensed to make oceans. Earth is a revolving ball with a molten core and a rocky and water covered surface.

Earth is one of eight planets along with minor planets and lesser moons, thousands of asteroids, billions of comets, uncounted meteorites, dust and gases orbiting the central star which is the Sun in our solar system. The planets are Mercury, Venus, Earth, Mars, Jupiter, Saturn, Uranus, and Neptune. Pluto has relatively low mass, and recently it has been reclassified to a minor planet. Dwarf planets include Ceres, Charon and Eris.

As mentioned, accumulating mini planets, dust and gases formed the Earth about 4.6 billion years ago. Compression caused by gravity produced immense internal heat and pressure. The Earth's molten surface cooled and hardened, but its interior is intensely hot. Oceans cover the majority of the Earth.

In space, Earth is a dense rocky planet, third closest to the Sun, and small compared with Jupiter and Saturn. While Earth tilts on an angle of 23.5 degrees and it rotates on its axis each day of 24 hours, it also orbits the Sun each year of 365 days, held in orbit by the Sun's gravity. From space the Earth looks blue and calm, but under its oceans, deep beneath the crust, the Earth's core is fiery and white hot.

An atmosphere of invisible gases mainly comprising nitrogen and oxygen surround the Earth. The crust of the rocky outer surface of Earth is about 6 km thick under oceans and up to 64 km thick under mountain ranges. The crust consists largely of granite under continents and basalt below oceans. The semi-molten mantle flows sluggishly between the crust and outer core. The dense molten outer core may be mainly iron and nickel with some silicon. The inner core has intense pressure and high temperatures of 3700 degrees C. From the Earth's core, convection currents convey heat through the mantle to the crust.

The Earth has a diameter of 12,756 km. It has an equatorial circumference of 40,075 kilometres. The Earth has a mass of 5976 million million million tones. The Earth has a surface area of 510 billion square kilometres. The Earth is located 150 million km from the Sun. The Earth orbits the Sun at 29.8 km/sec. Oceans cover 71% of the Earth's crust. Earth is a symmetric sphere because it bulges in the middle.

The most abundant elements in the Earth's crust are oxygen (47%), silicon (28%), aluminum (8%), iron (5%), calcium (4%), sodium (3%), potassium (3%), magnesium (2%), and other elements. Heavier metals such as iron and nickel are formed in the core.

Gravity anomalies reinforce the theory of isostasy which describes a state of balance in the Earth's crust where continents of light material float on a denser substance into which continental roots project like the underwater mass of floating icebergs.

The Earth's crust is a floating mattress. It is a restless jigsaw puzzle of oceanic and continental plates coupled to a rigid slab of upper mantle. Heat rising from the Earth's core and lower mantle are believed to produce convection currents that produce plate tectonics. Constructive margins are sub oceanic spreading ridges formed between two separating oceanic plates. Destructive margins are oceanic trenches where an oceanic plate dives below a less dense continental plate. Conservative margins are two plates sliding past each other. Active margins are where colliding continental and oceanic plates produce volcanic eruptions, earthquakes and mountain building.

The surface of the Earth's crust receives significantly more energy from the Sun above than the hot molten core below. Sunshine warms the tropics more than the polar regions. This uneven heating creates belts of differing atmospheric pressure. Winds blow from high to low pressure. Winds drive ocean waves and surface currents spreading heat more evenly around the world. The Sun drives the water cycle. The resulting rain, rivers, ocean waves and glaciers sculpt the surface of the land. Gravitational energy provided by the Sun and Moon produces our ocean tides, and tidal energy inside the molten layers of the Earth.

The Earth's crust is the rocky surface layer which scientists and engineers know best. Rocks are mixtures of minerals. Most rocks consist of interlocking grains or crystals cemented together naturally. Rocks vary greatly in size, shape and mineral proportions. Geologists have identified the three main rock groups as igneous, sedimentary and metamorphic.

Soil forms as weathering breaks rock into particles ranging in size from clay, silt, sand, gravel, cobbles and boulders. Air and water fill gaps between the soil particles. Chemical changes help bacteria, fungi and plants to move in. The chief influence on overburden soils is climate.

From space the presence of mankind on earth is analogous to how humans view an ant hill. Our lives and infrastructure activities are a very tiny speck in the vast spectrum of geological time and space.

We know that winds blow and oceans flow, and these are not the only parts of the Earth that are dynamic. Terra firma or solid Earth is not solid, and it is not forever fixed on the world map in space and time. Land moves about in response to natural forces. The drift of continents has a major influence on our climate and our life.

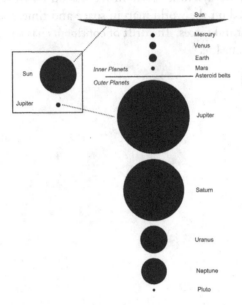

2.1 The Relative Sizes of the Sun and its Planets

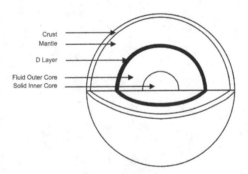

2.2 The Interior Structure of the Earth

Chapter 3
GEOLOGY

Geology is the wide scope of geoscientific investigation and observation which studies the composition and arrangement of the Earth's crust, followed by the application of these results to the art and practice of civil engineering, mining and geoenvironmental sciences.

Geologists recognize only one naturally occurring earth material called rock. Geoengineering differentiates between rocks and soils, and the reactions of rocks and soils to the forces imposed on them by earthworks and structures, and by groundwater chemistry.

The Earth has been undergoing changes for millions of years. Earthquakes, volcanic action, glaciers, floods, variable erosion and deposition, wind action and climatic factors produce the soil mantle on the Earth's crust, which is derived from the underlying rock.

Billions of years ago when the Earth cooled from its gaseous state, it was composed of molten material. Further cooling formed a crust at the surface. Hot molten material or magma still exists at great depth below the Earth's crust.

Continental Drift: Plate Tectonics

In 1915, Alfred Wegener, a German meteorologist and geophysicist, published The Origin of Continents and Oceans which set forth a radical hypothesis of continental drift.

Wegener suggested that a supercontinent named Pangaea, meaning all land, once existed. It was hypothesized that, about 200 million years ago, this supercontinent began breaking into smaller continents, which then drifted to their present positions.

Scientists collected substantial evidence to support the claims. The fit of South America and Africa, and the geographic distribution of fossils, rock types and structures, and ancient climates all seemed to support the idea that these current separate landmasses were once joined.

A current holistic theory to continental drift is plate tectonics. It is the framework for most geologic processes.

According to the plate tectonics model, the uppermost mantle along with the overlying crust, behaves as a strong, rigid layer, known as the lithosphere. This outermost shell overlies a weaker region in the mantle known as the asthenosphere. The lithosphere is broken into numerous segments called plates, which are in motion and are continually changing in shape and size. Seven major plates are recognized as the North American, South American, Pacific, African, Eurasian, Australian, and Antarctic plates. The largest is the Pacific plate, which is located mostly within the ocean and the continent moves with the ocean floor.

Intermediate-sized plates include the Caribbean, Nazca, Philippine, Arabian, Cocs, and Scotia plates.

The lithospheric plates move at very slow, but continuous, rates of a few centimetres a year. This movement is ultimately driven by the unequal distribution of heat within Earth. The titanic grinding movements of Earth's lithospheric plates generate earthquakes, create volcanoes, and deform large masses of rock into mountains.

Folds and Faults

For anticline folds the beds are convex upwards, whereas, for syncline folds the beds are concave upwards.

A fault represents a surface of discontinuity along which the strata on either side have been displaced relative to each other. The displacement varies from tens of millimetres to hundreds of kilometres. The fracture may be clean break or a widely developed, infilled or mineralized fault zone.

Discontinuities

A discontinuity represents a plane of weakness within a rock mass across which the rock material is structurally discontinuous. Although discontinuities are not necessarily planes of separation, most in fact are and they possess little or no tensile strength. Discontinuities vary in size from small fissures to huge faults. The most common discontinuities are joints and bedding planes. Other important discontinuities are planes of cleavage and schistosity, fissures and faults.

According to their origin, rocks are divided into three groups.

Igneous rocks are formed from magma, which has cooled and hardened slowly.

Sedimentary rocks are formed from weathering and disintegration of igneous rocks. The further weathering of fine rock fragments into sand, silt and clay and the deposition of these soils into depressions of the Earth's crust, followed by consolidation, cementation and pressure formed sedimentary rocks such as sandstone, limestone and shale.

Metamorphic rocks result from changes in heat or pressure, which alter other igneous or sedimentary rocks. Heat leads to recrystallization of the original rocks and pressure results in reorientation of the crystals and produces a banded or folded effect. Typical examples of metamorphism are sandstone to quartzite, limestone to marble, and shale to slate or mica.

A geologic survey of a region indicates the general type and pattern of subsurface strata to be anticipated. A geological site characterization is a preliminary step for conceptual design and construction purposes. A detailed geologic survey of a project site identifies the structural geology including the specific types of overburden soil and bedrock strata which are likely to be encountered. The geoengineering consultant may carry out a study of the regional geology and the local geology in order to plan and execute a major site exploration program. A geological site characterization is inadequate for quantitative engineering design and construction purposes.

Glacial Geology

Much of North America and Northern Eurasia are glaciated. A glacier is a large mass of ice which is formed by the accumulation and pressurized compaction of snow. Glaciers flow under the influence of gravity and/or extrusion. The effects of the glacial era are found as follows:

Glacial Advance: If the temperature trend is increasing, then the glacial winter has an advancing ice front. A ground moraine or glacial till sheet is formed when materials are eroded by the flowing glacier, transported beneath the ice and redeposited as a compacted unsorted mixture of boulders, gravels and rock dust or soil. A drumlin is one basic landform molded under the advancing ice and containing glacial till material in an oval ridge. Moraine tills are formed under advancing glaciers.

As the temperature drops, the advance of the ice front is halted temporarily to cause melting. Till material carried forward by the ice is dumped in a series of hills along the front of the ice called moraines. During temperature fluctuations if the ice advances and retreats several times, a succession of small ridges is formed, known as recessional moraines. When the ice melts without advancing again a fairly level ablation till is left. This consists of boulders, cobbles, gravel, sand, silt and clay all intermixed.

Glacial Retreat: When the temperature trend is decreasing and as the glacial summers become longer, the sun melts more ice than is replaced during the winters. The net effect is that the outer edges of the ice mass as well as the top of the glacier melt. The runoff from these melting periods would form rivers streaming from the ice mass. The melt water runoff forms the land by erosion and by depositing solids further downstream.

The glacial melt water produces streams and rivers. Assorted materials deposited in flat plains beyond the ice perimeter are called outwash plains. When melt waters are contained within valleys walls, the higher velocity flow carries heavy coarse particles and deposits them downstream. Streams flowing in tunnels beneath the melting ice blocks may have removed silt and clay and left long ridges containing coarse sand and gravel materials called eskers. Melt waters flowing variably along the edge of the glacier deposit gravel, sand and silt against the ice front. When the ice finally melted, these mounds containing alternate layers of silt and clay remained. They are known as kame moraines. Lake bottom deposits containing alternate layers of silt and clay are known as varved clays. Where the glacial river is emptied into lakes, broad fan shaped deltas were formed. These sediments grade rapidly from coarse to fine, and consist of gravel, sand and silt.

Glacial Erosion and Deposition

In upper North America, and for example in Central Canada we are especially interested in glaciers as the entire province of Ontario has been covered by at least two major ice sheets. The last glacial era in known as the Pleistocene Epoch. During this period, a known minimum of four advances and retreats of glaciers have occurred in Ontario. The most recent ice age is believed to have ended about 20,000 years ago. This most recent ice blanket reached into the south of Lake Ontario and Lake Erie.

The Wisconsin glacier advanced from two main centres in Labrador and west on Hudson Bay. The glacier advanced as far south as Ohio in the United States. The direction of flow is determined by the landforms and striations left by the glacier. As the ice advanced, the drainage of the Great Lakes region was completely disrupted by the damming of the natural water flow to the east. The water level rose in the Great Lakes until it found outlets to the south through the Mississippi and Mohawk River systems. After the recession of the glacier, the area around the Ottawa basin rebounded to its present elevation and water spilled down through the St. Lawrence River system.

Continental glaciers, because of their great size, tend to bury completely the area which they traversed. The weight of ice of 3000 m thickness in the centre of the glacier exerts an immense pressure of about 30 MPa on the lowest layer of ice, causing glacial flow. Glaciers erode rock by abrasion or plucking. Generally, the ice mass gouges out the softer materials and leaves harder rock masses standing. Since the softer rocks are more easily gouged out by the ice, they tend to form basins, which form lakes. In North America the fresh water Great Lakes are an example of this.

As the glacial summers become longer, the sun melts more ice than is replaced during the winters. The net effect is that the outer edges of the ice mass as well as the top of the glacier melts. The runoff from these melting periods would form rivers streaming from the ice mass. The melt water runoff forms the land by erosion and by depositing solids further downstream.

Where the geoengineering consultant can correlate topographical features to glacial deposits as described above, then favorable aggregate sources ranging to problematic subsurface deposits may be anticipated or avoided, respectively.

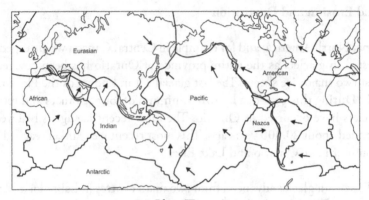

3.1 Plate Tectonics
Earth's Major Crustal Plates, with arrows to show
relative movement; mostly few cm/year.

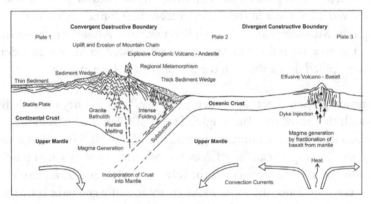

3.2 Plate Boundary Process

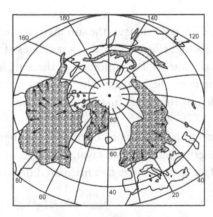

3.3 Pleistocene Glaciations in the North Hemisphere (E. Antevs)

Chapter 4

ROCKS AND MINERALS

Rocks and minerals are fundamental to the study of natural history, leading to an appreciation of nature and earth sciences. Geologists recognize only one naturally occurring earth material called rock. Engineers differentiate between rocks and soils. Rocks are generally bonded or cemented together and an initial yield resistance must be overcome mostly by high forces before they shear in an unconfined state. Soils exhibit very low resistance to shear in an unconfined state, and a very small force is required to cause soil breakdown. If rock is massive, then soil is considered to be a particulate system. Rocks and soils may be layered, but rock masses are fissured and jointed, and this means that rock masses are often controlled more in their reaction to forces acting on discontinuities by the discrete nature of the large fissured mass than by the properties of the small material samples in laboratory testing.

In an early chapter rock types are classified into three major groups: igneous, sedimentary and metamorphic.

Igneous Rocks

A. **Formations**

    Extrusive igneous rocks are formed in volcanic eruptions at surface.

    Intrusive or plutonic igneous rocks are produced within the earth.

    Magma
    *   Consists mainly of silicate materials
    *   Contains gases, such as water vapor

- Originates from as deep as 200 km
- Differs in rate of cooling, chemical composition, and amount of gases
- Produces intrusive plutonic igneous rocks.

Lava
- Similar to magma
- Most gaseous elements have escaped
- Produces extrusive volcanic igneous rocks

B.   Structure

Crystallization (ordered pattern of ions)
- Rate of cooling strongly affects crystal size
- Slow cooling = large crystals
- Rapid cooling = small crystals

Amorphous (unordered pattern of ions)
- Occurs with instant cooling or quenching
- Produces rocks referred to as glass

C.   Classifying Igneous Rocks

Texture, size and arrangement of interlocking crystals
- **Fine grained**
  - i) Form at surface or within upper crust
  - ii) Openings left by gas bubbles are called vesicles
- **Coarse grained**: Formed deep within the interior
- **Porphyritic**: Large crystals imbedded in matrix of smaller crystals

By mineral composition
- Depends on composition of magma and cooling
- Same magma can create rocks of varying mineral content
- As magma cools, certain minerals crystallize first at very high temperatures

METAMORPHIC ROCKS

A.   Formation

Metamorphic: To change form

Regional metamorphism: Material under intense stresses and high temperatures

Contact (thermal) metamorphism:

Changes caused by proximity to magma

Low grade metamorphism: Shale becomes slate
High grade metamorphism:
Often form during mountain building, fossils become rock, most metamorphic rocks are harder than sedimentary rocks
Metamorphism changes texture
Low-grade metamorphism makes material more compact, more dense
Foliated texture: Particles of material are brought into line with one another

B. **Agents of Metamorphism**
Heat
- Most important agent
- Provides energy for chemical reactions
- For example: Clay recrystallizes into a mineral at great temperature

Pressure
Chemical Activity: Most common chemical agent is water

C. **Types of Metamorphic Rocks**
Hornfels: Fine grained, dark flinty rock with randomly arranged minerals
Slate: Fine grained, often gray, foliated rock split easily along cleavage, planes of mica flakes aligned by pressure
Marble: Granular or sugary-textured rock formed from limestone
Phyllite: Silky, foliated rock more coarsely grained than slate
Schist: Foliated rock, more coarsely grained and of higher metamorphic grade than phyllite
Gneiss: Foliated, banded rock of the highest metamorphic grade
Quartzite: Very hard, granular quartz rock, formed from sandstone

SEDIMENTARY ROCKS

A. **Formation**
Mechanical Weathering forms sediments
- Frost wedging
- Unloading
- Biological activity: roots, burrows, bombs
Chemical Weathering forms sediments

- Water to rust (oxidation)
- CO_2 and water make carbonic acid
- Granite reacts with carbonic acid to make clay minerals, potassium and silica

Lithification: The process of sediments becoming solids

- Compaction: weight compresses deeper sediments
- Cementation: materials carried in solution cement particles together

B. Classifying Sedimentary Rocks

Detrital Sedimentary Rocks

- Accumulated debris from weathering and transport
- Made up of mostly clay minerals and quartz
- Differentiated by particle size
- Conglomerate: made up of gravel-sized particles
    i) Well sorted
    ii) Poorly sorted
    iii) Close packed
    iv) Loosely packed
    v) Imbricated (overlapping or tiled)
    vi) Graded bedding: Smaller stones on top, larger on bottom
- Breccia: made up of angular particles
- Sandstone: made up of sand-sized pieces
- Shale: very fine grained

Chemical Sedimentary Rocks

- Formed from materials in solution in bodies of water
- Most abundant form is limestone
- Coal, although made from organic matter, is considered part of this group.

C. Features of Sedimentary Rocks

- Sedimentary rock forms at surface
- 5% by volume of Earth's crust, but much of outer surface
- It is from sedimentary rocks that we can reconstruct much of Earth's physical history
- Keys to past environments, plants, animals, and asteroids
- Petroleum and natural gas occur in pockets of sedimentary rocks

MINERALS

A mineral is a naturally occurring, inorganic, solid material with a definite atomic structure.

A. Properties of Minerals
Luster: appearance or quality of light from surface
Colour: nature of light and cause of colour
Streak: colour of material in powdered form
Shape is determined by cleavage, crystal form and fracture
Specific Gravity
- The ratio of the weight of a mineral to the weight of an equal volume of water
- Density of water = 1 gm/cm^3= 1 gm/ml
- Density of lead = 7.7, aluminum = 2.7, osmium = 22

B. Mineral Groups
Silicates: minerals with silicone and oxygen
- **Silicon** is a semi-metal which forms a pyramid-shaped structure with oxygen
- Other forms include single tetrahedron(olivine), chains (augite), sheets (micas), 3-D structures (feldspars and quartz)
- Non-Silicates
  i) Make up one-fourth of continental crust
  ii) Carbonates: minerals with carbon and oxygen including calcite, from which we procure limestone (roads) and marble (decorative slabs)
  iii) **Oxides**: oxygen based solids, i.e., ice
  iv) Sulfides (S), sulfates (SO4) halides (CL,F)
  v) **Halite**: mineral form of salt
  vi) **Gypsum**: plaster, calcium
  vii) **Native metals**: iron, zinc, gold, silver, nickel

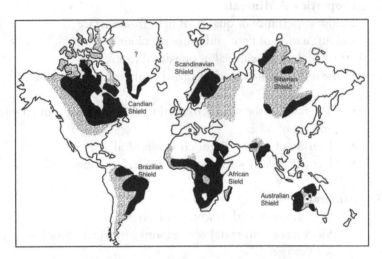

4.1 Precambrian Shield Areas of the World
Rocks exposed at the surface are shown in black
Platform areas where Precambrian is covered
by shallow sediments is stippled

Chapter 5

HYDROLOGY AND THE HYDROLOGIC CYCLE

Water

The most common and most significant fluid earth material that the engineering geologist must deal with is water. The ever changing migration of atmospheric, surface and ground water is a complex interdependent system called the hydrologic cycle. Although the hydrogeologist is concerned chiefly with groundwater, the hydrologist is concerned mainly with surface water. The aspects of the hydrologic cycle should be understood before an accurate characterization of the subsurface portion of the cycle can be achieved.

- The Earth contains 326 million cubic miles of water
- 97% of all Earth's water is in the oceans
- Another 2% is in glaciers and ice sheets
- 1% is in lakes, streams, subterranean water and atmosphere
- 90% of all the ice on Earth, and 70% of all its fresh water, is in Antarctica
- The largest single body of fresh water is the Caspian Sea, located in Asia
- The largest fresh water lake on the planet is Lake Baikal, located in Russia

Physically, water, unlike any other common earth material, can exist in three forms: solid (ice), liquid (water), and gas (vapour) at ambient temperatures and pressures. Moreover, water can occur in these three phases at the same temperature, 0°C at atmospheric pressure. Another unique feature of water is that it reaches its maximum density at 4°C while in the liquid phase (ice at 0°C floats in water) and expands in volume by about 10% as it freezes. Because water has unique physical characteristics at ambient temperatures and pressures on the earth, it is cycled through the environment. The hydrologic cycle is the circulation of water from the ocean (fluid), to the atmosphere (vapor), to the land (rainfall), and back to the ocean (runoff). The hydrologic cycle can be divided into a series of processes: evaporation and transpiration, precipitation, and runoff and infiltration, which are all related through the hydrologic equation:

Precipitation *(P)* = Runoff *(Q)* + Evaporation and Transpiration *(ET)* + Infiltration *(I)* + Storage *(S)*

This equation states that water cannot be created or destroyed, and therefore the hydrologic cycle must balance. Calculating this balance, is a complex scientific problem.

The most obvious part of the hydrological cycle involves rainfall that runs downhill and accumulates into streams and rivers that combine and eventually reach the sea. Less evident is the evaporation that simultaneously takes place, rises to form clouds, and returns to the ground surface as rain or snow to rejoin the hydrological cycle. Even less evident is water that infiltrates into the soil where its influences are manifest in springs, wells, and green lawns. These factors plus the influence of transpiration from vegetation and recycling by animals complete the hydrological cycle.

The percentage of precipitation that infiltrates into the ground is close to 100% in cavernous limestone areas or desert sands and decreases to 10 to 20% on impervious soils such as in shale or granite. Much depends on the rate of rainfall and the moisture condition of the soil.

Runoff water is the main sculptor of landscape, and erosion patterns and the resistance to erosion are important clues to the compositions of rocks and soils. The management of infiltrating water is a major concern for geotechnical engineers as it influences soil strength, reservoir levels, and soil conditions and erosion at construction sites. Water can reduce the bearing capacity of soils and is a key factor in landslides. Geotechnical engineers also are concerned with seepage of water through earth dams, levees, and soil lining irrigation ditches, as well as flow into underdrains and wells. It is important that an engineer has a working knowledge of the principles governing the flow and retention of water in soil, and the effect of water on strength and stability of the subsurface materials.

The Water Cycle

The Path Water Takes

- The Earth's water evaporates due to the heat and energy from the Sun.
- Wind transports the evaporated water to regions where clouds form.
- Rain falls, returning the water to the Earth.
- The rainwater then makes its way back to places on land and sea where it evaporates again, starting the process anew.

Water That Falls on Land

- Water that falls on land is either absorbed by plant life or falls to the ground.
- Water that is absorbed by plants is later released into the atmosphere through a process called transpiration.
- Crops yield an average of 600 mm of transpiration per year.
- Trees yield twice as much.
- Water not absorbed by plants moves downward into the earth, then into lakes, streams and oceans.

Water Cycle Dynamics

- The water cycle is a dynamic equilibrium. This means that the total amount of water contained in the atmosphere remains essentially the same at all times, whether derived from land or water.
- On land, precipitation exceeds evaporation. This means that in an average period of time, more rain will fall on an area than can evaporate from that area in the same amount of time.
- On water, however, evaporation exceeds precipitation.
- Thus, the dynamic water cycle equilibrium is maintained.

SURFACE WATER SHAPING THE EARTH

Moving Water

- Moving water is the single most important factor in shaping the Earth's surface.
- Water is set in motion by gravity.
- Liquid friction, or erosion, is the action of water constantly rubbing against rock, forms channels through the rock. The amount of liquid friction in any given spot defines the shape and roughness of the channel.
- A gradient is a vertical drop over a defined lateral distance.

Energy and Discharge

- The steeper the gradient, the more potential energy of motion that is released.
  - i) The law of conservation of energy states that energy cannot be either created or destroyed, but is conserved at all times.
  - ii) Potential energy lies dormant while awaiting a force that will convert it to kinetic energy.
  - iii) Gravity releases the potential energy that causes water to flow.
- Discharge is defined as the volume of water that passes in a given unit of time, e.g. cubic metres per second.
- Discharge is not a constant and can be affected by numerous factors such as gravity, gradient, flow channel characteristics, etc.

Profile

- Consider the cross sectional view of a stream from source to mouth.
- Water usually flows faster at the head, the source of the stream, than at the mouth, the opening where the stream or lake, river, etc. opens onto another body.
- The steeper the gradient, the smaller the discharge.
- The smaller the gradient, the larger the discharge.

Base Level

- Streams cannot erode their channels endlessly. The deepest level to which they can erode is called their base level. At this level, so much potential energy has been released that there remains too little to be of consequence.
- Sea level is the ultimate base level since it contains no potential energy.
- Temporary or local base levels are responsible for lakes and larger streams which can, under rate circumstances, become deeper.

Dynamic Process

- The Dynamic Process: The deeper a channel becomes cut, the less energy the water has, therefore it moves slower and moves less silt.
- A lake, for example, prevents its feeder streams from eroding below its level at any point upstream from the lake.
- Any change in base level will cause a corresponding change in stream activity.

Work of Stream Channels

- Streams are any steady current of water, whether in a brook or an ocean.
- Their constant motion produces erosion, the removal of rocks and soil from a given area.
- These are then transported by the stream in three ways:
  - i)   In solution, with the salt dissolved into the water of the stream
  - ii) In suspension, which remains separate and differentiates between rock, soil and salt
  - iii) As sediment which rolls along the bottom and is carried and deposited in other areas, by the stream. This sediment is called bed load.

Water and Velocity and Erosion

- As a stream channel matures, it has less of a gradient.
- This leads to increased velocity in the outer channel water flow, thus increasing the erosion of outer banks.
- The proportionately decreased velocity on the inner curve results in increased deposition, which is called a point bar.

- There is an increased velocity on the downstream side of a meander caused by gradient. This, in turn may cause bends to migrate downstream.

Drainage Patterns

There are four major types of drainage patterns.

Deadritic

The stream system resembles the branching of a tree.

Radial

Streams run in all directions from a central high point.

Rectangular

The pattern makes frequent right angled turns. Usually develops on fractured or jointed bedrock.

Trellis

When tributaries of an almost parallel structure occupy valleys formed by bent rock layers deformed from an originally horizontal plain, called folded strata.

Water is a commodity whose value varies according to locality, purpose and circumstance. Water is not evenly distributed. Nine countries account for 60% of all available fresh water supplies and among them only Brazil, Canada, Colombia, Congo, Indonesia and Russia have an abundance. America is relatively sufficient, however China and India, with over a third of the world's population between them, have less than 10% of its water.

According to the World Bank 2010 reports the earth's population has grown from 2.5 billion in 1950, to 6 billion in 2000, nearly 7 billion in 2010, and it is projected to increase to 9 billion in 2050.

The global water resources include 97.5% from salt water oceans and 2.5% from fresh water. Fresh water comprises 68.7% from glaciers and ice cups,

30.1% from groundwater, 0.8% from permafrost and only 0.4% from surface water and the atmosphere.

Water abstraction from rivers, lakes and groundwater goes 67% to agriculture, 20% to domestic and industrial, 10% to power generation and 3% to evaporation from reservoirs.

Throughout history, human dependence on water has made us to live near it or organize access to it. Water makes up about 60% of the body. Water provides life and food, a means of transport, a way of keeping clean, a mechanism for removing sewage, a habitat for fish and animals, a medium to cook, to swim, to skate and sail, a place of beauty to provide aspiration. Water is the new oil, a resource long squandered, now growing expensive, and soon to be overwhelmed by insatiable demand, and in some cases wars may occur between countries claiming rights to surface water resources particularly lakes, rivers and dams.

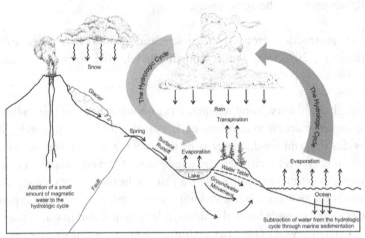

5.1 Hydrologic Cycle (Davis and DeWiest, 1966)

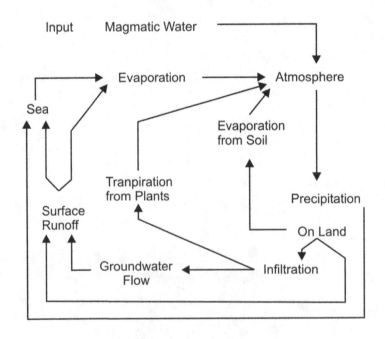

5.2 Hydrologic Cycle

Chapter 6
GEOTECHNICS

Geotechnics as an applied science is the overlap between the geotechnical engineering aspects of civil engineering (geotechnical, structural, environmental, transportation, water resources, construction, surveying and mapping), and, the engineering geology aspects of geology (engineering geology, hydrogeology, geophysics, geochemistry, seismology, geomorphology, paleontology, petrology, and economic geology).

Geomechanics is concerned with the mechanical responses of all geological materials, mainly rock and soil. Mineral soil particles result from weathering of the rock that forms the crust of the Earth. The geoengineering mechanics problem faced in all subsurface design is the prediction of the performance of the substructure under the loads imposed on it during its prescribed functional operation.

The judgement of the civil or mining engineer supported by the skilled observations and the detailed analyses of the specialist geoengineering team make great contributions to successful projects.

A full appreciation of the limits of any investigation and of the subsurface conditions comprising the soil, rock, groundwater and geochemical conditions is paramount. The engineer must be able to visualize three dimensional subsurface conditions in spatial terms, recognizing the somewhat static nature of soil and rock, and the mostly dynamic and complex migratory nature of groundwater and contaminants. In part of the hydrologic cycle there is a crossover from surface water to groundwater, and the relationship of both to the weather must always be remembered, recognizing the significant differences between dry and wet weather, cold and hot temperatures, and on pervious and impervious ground conditions. The geoengineering study of any project site often includes a summary of the local weather conditions from past to present. Weather conditions may be continued and reported in daily construction site reports.

In Ottawa there are recent occurrences of soil dehydration attributed to climate change that have resulted in unanticipated soil consolidation and soil settlement of weak marine clays (including Leda Clay) causing significant cracking of low rise buildings founded at shallow depths on the clay.

There are ongoing examples of the profound effects of unfavourable geoengineering conditions on civil engineering and mining works. This indicates how closely the science and art of geoengineering are related and how dependent surface and subsurface structures must be upon geoengineering practices.

For the owner, designer, contractor or regulatory agency engaged in the planning, design, construction, operation, maintenance or management of civil works and mining projects, the properties of soil, rock, groundwater and contaminants, and their impact on earthworks, structures and humans is of primary importance. These earth engineering properties include structure, strength, deformation, permeability, surface water and groundwater interaction, geochemistry and seismicity.

Unlike other homogeneous and isotropic construction materials such as steel, concrete, timber, plastic, etc. we cannot standardize ground conditions because they are nonhomogeneous and anisotropic, and wide variations in the soil, rock and groundwater layers, and the geochemistry, may occur between the testhole locations over a project site.

The design and construction of all earthworks and structures deserves a subsurface site investigation and related geoengineering recommendations. A comprehensive site assessment report is a good investment – it buys knowledge, insurance, economy in design and construction, and it reduces claims.

The project team members should read and implement the specialist geoengineering reports on geology, geotechnical and geoenvironmental engineering, hydrogeology, hydrology, and geochemistry, and incorporate the findings into the planning and feasibility studies, and the recommendations into the detailed design, construction drawings and project specifications. The spills and containment and remediation of contaminants are highly dependent on knowing the hydrogeology and geochemistry of the site.

The project manager should ensure that all geoengineering requirements for the project are carried out, and seek geoengineering assistance throughout the project including fulltime onsite supervision for earthworks, slopes, foundations, excavation, backfill, groundwater seepage, erosion, drainage, contaminant migration and site remediation.

The important effects of geoengineering conditions and practices on major civil engineering and mining works are seen in the underground tube railways in London and New York, the underground mines in Timmins, the major Hoover and Aswan Dams, and other important superstructures and substructures worldwide.

In London, the city is built on a large basin of unconsolidated and weaker soil know as London clay, therefore underground tube railways are tunneled far below ground level and constructed in clay that was easily and economically excavated, however the same clay must be carefully investigated and foundation designs going down to deep high strength strata are often required for tall and heavily loaded buildings. In New York, by comparison, the surface of Manhattan Island on which the city is underlain mostly by schist bedrock. The underground railways had to be built in carefully excavated rock cuts slightly below the ground surface level necessitating more significant equipment, time and costs than for unconsolidated overburden soil. Conversely the shallow bedrock conditions in New York are favourable for the foundation support of high rise buildings. The once deepest copper sulfide mine in Timmins excavated to about 3 km, (10,000 ft) exhibited hard igneous bedrock conditions, however when the inclined rock joints were relaxed during open pit and underground mining operations, then additional underground support was needed to prevent massive rock wedge movements.

The Hoover Dam on the Colorado River in the United States and many other successful dams in the world show the value of accurately knowing beforehand the subsurface conditions obtained from geologic, geophysical and geotechnical investigations, as do large and small bridges around the world. The foundation engineering for the great bridges of New York, San Francisco, PEI Confederation Bridge, and European and Asian sites are particularly notable.

Whether the geoengineering specialist gains an appreciation of subsurface conditions by training or intuition, or both, that appreciation is very important for the field investigations, office designs, construction approaches, mining methods and earthworks management.

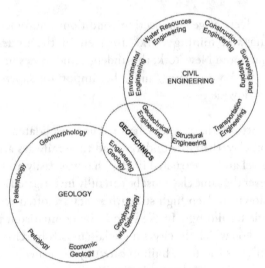

6.1 Geotechnics:
Application of Geologic Principles to Engineering Works

Chapter 7
SUBSURFACE INVESTIGATION

The successful design, construction and performance of an engineering structure or mine can be accomplished only when the character of the soil, rock and groundwater conditions on which or within which it is to be built, is known. For this knowledge to be obtained the ground must be carefully studied by insitu investigations. Engineering structures such as roads, dams, buildings, tunnels and underground works, are normally constructed according to the requirements of a specific design and from selected construction materials. By observing these requirements the strength of completed structures is known and their response to load and displacement may be predicted. To obtain comparable information about the soil and rock against which the structure will react it is necessary to survey the geological processes which formed the soils and rocks, to better understand their structural geology design, and to investigate the nature of the subsurface materials of which they are composed.

A subsurface investigation is an assessment of the subsurface conditions at selected points. It comprises data collection and data analysis using methods such as geological and geophysical surveys, boring and sampling, insitu testing, visual inspection, local experience, laboratory testing of subsurface samples; and long term groundwater level observations using sealed standpipe piezometers and monitoring wells. Various mechanical, electromagnetic, electrical resistivity and seismic devices using rapid data acquisition units with analog and digital readout are used in field instrumentation and monitoring programs.

The subsurface investigation is the first and most important step for all earthworks and structures. A borehole drilling and sampling investigation should be carried out for all earthworks and structures, even modest ones, (i) for feasibility assessments (ii) before design is undertaken, (iii) before the approval of a regulatory body is sought, and (iv) before construction tendering commences.

It is important that subsurface investigations be carried out under the direction of geoengineering consultants and personnel with knowledge and experience in planning and executing detailed drilling and sampling methods. It is important that drilling crews be experienced in soil and rock borings for subsurface explorations. The drilling equipment comprises rotary and percussion advance methods using truck, track and barge mounted carriers, to determine:

a)   the nature and sequence of all stratigraphic units,
b)   the varying groundwater conditions at the site,
c)   surface water influences,
d)   the physical properties of the soils and rock underlying the site,
e)   the mechanical properties such as strength, compressibility and permeability of the different soil or rock strata, and
f)   the geochemical composition of the soil, rock and groundwater,
g)   other specific information, when needed, such as the characteristics of all surrounding surface and subsurface structures, and underground openings, both temporary and permanent,

Site investigations should be organized such that all possible information is obtained to provide a thorough understanding of the subsurface conditions and the probable behaviour of the existing and proposed structures.

Particular attention should be paid to our present approach in a subsurface investigation, whereby we examine a very small sample versus the very large and varying earth mass and dynamic groundwater movements on site, and subsequently attempt to predict the earth and groundwater behaviour for the proposed structure.

This subsurface investigation approach by the geoengineering consultant requires knowledge, experience, sound judgement, and some intuition and favourable luck.

The field engineering inspector, at the time of construction, has the benefit of visualizing the soil, rock, groundwater and observable contaminants instead of only the sample obtained in a borehole investigation.

All subsurface variations from the predicted subsurface conditions should be reported immediately by the field engineering personnel to the specialist geoengineering consultant for verification and follow up recommendations, as work progresses.

Geotechnical Field Testing Methods

i)   Standard Penetration Test

The standard penetration test (SPT) based on ASTM D1586 is carried out by driving a standard split spoon sampler of fixed dimensions under a constant driving energy, usually a weight dropped a uniform number of times, over a uniform distance. A small diameter split spoon sampler is designed to extract a disturbed sample.

ii)   Dynamic Cone Penetration Test

The dynamic cone penetration test (DCPT) uses a rod with a standard conical end driven into the soil with known energy as with the split spoon test. No sample is available. The driving resistance provides relative density and soil strength information.

iii)   Static Cone Penetration Test

The static cone penetration test based on ASTM D3441 differs from the dynamic cone test since the energy is applied as a constant pressure rather than a series of regulated impacts. It measures strength and density in sand, and strength in clay.

iv)   Plate Bearing Test

The plate bearing test based on ASTM D1194 is carried out in an open pit with a steel plate loaded in equal increments. Settlement is measured after each load increment is added, allowing appropriate time intervals between load applications. It measures modulus and bearing capacity.

v)   Field Vane Test

The field vane test (FVT) based on ASTM D2573 are used in soft clays to determine in-place shear strength. The field vane usually consists of four blades at 90° to each other, which are attached to the end of a rod and inserted in the borehole and carefully pushed into the soft soil. The torque required to rotate the shaft is measured, from which the soil shear strength can be derived.

vi)   Pressure Meter Test

The pressure meter test (PMT) consists of a metal cylinder covered with an inflatable rubber membrane, which can be activated by gas or liquid pressure from a measured source. This allows the determination of the load and deformation characteristics of the soil or rock in open boreholes.

End effects are minimized by the use of the guard cells at each end of the test cylinder, which are expanded tightly to the wall of the borehole prior to testing.

vii)   Permeability Test

A permeability test involves the addition of water to soil or rock to determine the ability of the porous media to pass water through itself. Such tests include falling and constant head tests in cased or uncased boreholes. This includes basic percolation tests in shallow test holes as well as sophisticated full scale aquifer and pumping tests for hydrogeologic assessments.

Unlike other engineering disciplines the geotechnical engineer cannot undertake analysis and design by choosing the required proportions and properties of any construction material available. The ground is part of nature and man has had no control over its character. It must be investigated.

An accurate, complete and unclear picture of the site conditions must be obtained. Skimping on a site investigation is for fools looking for serious trouble.

A site investigation should be phased into a desk study, site reconnaissance and then a subsurface drilling investigation. In view of our industrial heritage it is increasingly important to undertake a chemical assessment either prior to or alongside the subsurface investigation.

It is emphasized that the methods of exploration, procuring samples, conducting insitu and laboratory testing are not perfect. They cause disturbance to the soil structure, they change the stress state and do not always model appropriately the conditions applied by the engineering works. Their limitations should be appreciated, as expected from knowledgeable and experienced practitioners. New developments in site investigation techniques are progressing.

To prepare a thorough interpretative report the geoengineering specialist not only needs a good knowledge of geology, soil mechanics, rock mechanics, hydrogeology and geotechnical engineering with some chemistry, but must also have well developed communication and presentation skills to address site specific needs.

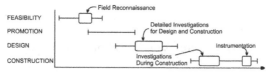

7.1 Example of Phased Ground Investigation
Rectangles represent periods of work on site (Kennard, 1979)

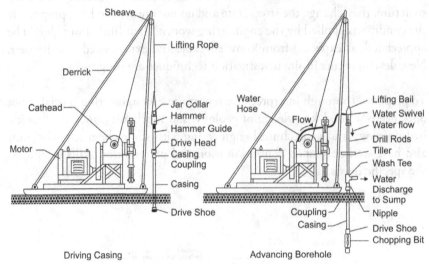

Sheave

Lifting Rope

Derrick

Cathead

Jar Collar
Hammer
Hammer Guide
Drive Head
Casing
Coupling

Casing

Drive Shoe

Motor

Water
Hose

Flow

Lifting Bail
Water Swivel
Water flow
Drill Rods
Tiller
Wash Tee
Water
Discharge
to Sump
Nipple
Drive Shoe
Chopping Bit

Coupling
Casing

Driving Casing   Advancing Borehole

7.2 Early Percussion Washboring Drill Rig

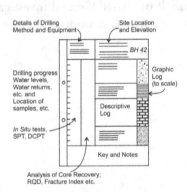

Details of Drilling
Method and Equipment

Site Location
and Elevation

BH 42

Drilling progress
Water levels,
Water returns,
etc. and
Location of
samples, etc.

Graphic
Log
(to scale)

Descriptive
Log

In Situ tests,
SPT, DCPT

Key and Notes

Analysis of Core Recovery;
RQD, Fracture Index etc.

7.3 Example of Geotechnical Borehole Log
SPT = standard penetration test
DCPT = dynamic cone penetration test
RQD = rock quality designation

Chapter 8

SOIL CLASSIFICATION

Soil is that portion of the Earth's crust, which is fragmentary, rock dust or such, that some individual particles of a dried sample may be readily separated by agitation in water. It includes boulders, cobbles, gravel, sand, silt, clay and organic matter.

The two more important tests used in classifying soils are grain size distribution to measure grain sizes, and plasticity to measure grain types.

For soil classification there are three major groups:

a) **Coarse Grained Soils** - particles which are large enough to be visible to the naked eye. They include gravels and sands, which are larger than the No. 200 sieve (0.075 mm). These are generally referred to as cohesionless or noncohesive soils.

b) **Fine Grained Soils** - cohesive silts and clays which are plastic.

c) **Organic Soils** - soils having a high natural organic content, which are fibrous and compressible.

Soils are identified using American Society for Testing and Materials (ASTM) methods to their particle size and distribution for coarse grained soils, and their liquid and plastic limits for fine grained soils. The soils are classified using standards such as the ISSMFE in Canada and the Unified Soil Classification System in the USA.

Visual Soil Classification

Generally, there is little problem in recognizing coarse grained soils, which comprise sand, gravel and cobbles. Some confusion occurs when distinguishing silts from clays. Silts are frequently classified incorrectly as clay. This error can lead to construction problems since the behaviour of silt, both in an excavation base and a construction backfill material, is significantly different from either clay or sand.

Based on a visual inspection of several samples taken from a site, it is important which soils are the coarse grained sands and gravel. Even fairly fine sands are easily recognized. The difference between silt and clay is not as obvious.

In the field, it is usual to distinguish between silts and clays using a dilatancy test. A wet 'pat' of soil is shaken in the palm of the hand and if the soil exhibits dilatancy, it will show free water on the surface. Moreover, this free water will disappear when the soil is squeezed and the soil will then become stiff and crumbly. Soils, which exhibit dilatancy, are normally silts. Clays will not exhibit dilatancy.

The sense of touch can be used to distinguish silts from clays and from sands, whereas sands have a gritty feel and the grains are visible to the naked eye. Silts have a rough, gritty texture and clay has a smooth plastic feel. Clay can be rolled into very thin threads without breaking. Clay sticks to the fingers and it dries slowly, but silt dries fairly quickly and can be dusted off leaving only a stain. It is not always possible to categorize soil as a strict sand, silt or clay type. Frequently, any glacially deposited material will consist of a combination of types, although normally one is predominant. For example, a silty clay is predominantly clay with some silt content, i.e. it will not exhibit dilatancy and will feel mainly smooth and plastic, however, with some evidence of a rough texture. On the other hand, a clayey silt will show dilatancy although it will also be plastic to some degree, i.e. it can be rolled in the fingers.

Soil Laboratory Tests

In addition to visually identifying soils in the field, it is important to carry out laboratory tests such as sieve and hydrometer analyses using ASTM

D422 for soil classification. Soil classification may be carried out on the basis of grain size diameter as follows:

Coarse Gravel	75 mm to 20 mm
Fine Gravel	6 mm to 2 mm
Coarse Sand	2 mm to 0.60 mm
Medium Sand	0.60 mm to 0.20 mm
Fine Sand	0.20 mm to 0.60 mm
Silt	0.60 mm to 0.002 mm
Clay	less than 0.002 mm

The above divisions are not exact. The characteristics of a soil are indicated not only by grain size but also by grain shape such as angular or rounded, water content, and grading such as well graded or uniformly graded.

It is important to be able to distinguish the differences between various soils such as clayey silt, sand and medium gravel. A gradation curve will provide an indication of the uniformity of the material. For example, a poorly graded or uniform sand is essentially one size, e.g., medium sand. The uniformity of soils significantly affects compaction and permeability.

Imported Granular Fills

There are many types of imported soils used in earthworks, construction and mining. Generally, the materials are granular or cohesionless since they are easier to excavate from borrow pits, easier to handle and compact, etc. For example, in Canada, Ontario OPSS 1010 specifies the gradation requirements for select granular fills. Depending on the function of the imported fill, material can vary from a Granular A to a sand cushion or Granular C. By definition, a Granular A material is a superior type of granular soil having closer specified tolerances regarding particle size than either a Granular B or Granular C. It should not be overlooked that Granular A, contains a specified small portion of crushed or angular stone. Granular A, therefore should be used in the top course of a road, or pipe bedding, since it has superior strength properties compared to the underlying granular subbase, Granular B.

Often, there is confusion as to the meaning of 20 mm crushed stone. Where specified as 20 mm clear crushed stone, it contains angular crushed particles of stone at 20 mm size. By definition, it is uniformly graded.

On the other hand, 20 mm crusher run stone contains angular crushed particles from 20 mm down to dust, i.e. well graded.

In most of Central Canada where there is a good supply of granular soils they are generally relatively clean, that is, free from clay and silt contamination, and they are reasonably well graded and not predominantly one size. For the most part, excavated sand and gravel from borrow pits are graded which enables these materials to be easily compacted at the optimum moisture content. Native borrow materials vary considerably from consolidated high strength glacial tills in the upper regions of North America and Europe; to unconsolidated alluviums and deltaic sediments in central Mediterranean regions; to loess, laterites sand cemented soils in arid tropical regions in equatorial parts of Africa and elsewhere.

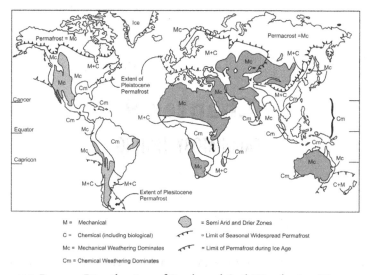

M = Mechanical

C = Chemical (including biological)

Mc = Mechanical Weathering Dominates

Cm = Chemical Weathering Dominates

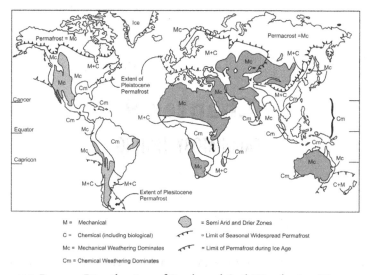 = Semi Arid and Drier Zones

= Limit of Seasonal Widespread Permafrost

= Limit of Permafrost during Ice Age

8.1 Present Distribution of Rock and Soil Weathering Types
Based on data from Washburn (1979); Meigs (1953); Flint (1957); and Strakhov (1967)

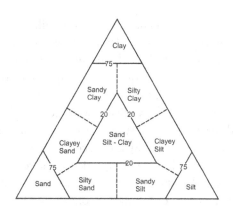

8.2 Soil Classification
Intermediate and Fine Grained Sediments

Chapter 9

LABORATORY TESTING

The soil, rock and groundwater samples recovered from subsurface exploration can be tested in the laboratory. The step by step standardized procedures for performing the laboratory tests are outlined by the American Society for Testing and Materials (ASTM) and various standards. The most common types of soil laboratory tests include:

a) basic index test for physical characteristics of soil.
b) oedometer apparatus to assess settlement or expansive soil behavior.
c) shear strength is a basic technical parameter measured by direct shear, unconfined and triaxial cell equipment as required for the analysis of earthworks, foundations and slope stability problems.
d) permeability is defined as the ability of water to flow through a saturated soil measured using the constant head or falling head laboratory permeameter to determine the hydraulic conductivity 'k' or the coefficient of permeability for the drainage properties of the soil.
e) compaction tests in the laboratory include Modified Proctor and Standard Proctor used to determine the relative compaction of fill.
f) pavement tests are for subgrade strength and aggregate quality.
g) geochemical properties for environmental assessment.

The common soil laboratory tests used in geoengineering are as follows:

Type of Condition	Soil Properties	Specification
Index tests	Water content (moisture content) Unit Weight Specific gravity Particle size (sieve and hydrometer) Atterberg limits Sand equivalent	ASTM D 2216-92 Block samples or sampling tubes ASTM D 854-92 ASTM D 422-90 ASTM D 4318-95 ASTM D 2419-95*
Permeability	Constant head Falling head	ASTM D 2434-94 ASTM D 5084-97 **
Settlement	Consolidation Collapse Organic Content Fill compaction: Standard Proctor Fill compaction: Modified Proctor	ASTM D 2435-96 ASTM D 5333-96 ** ASTM D 2974-95 ASTM D 698-91 ASTM D 1557-91
Expansive soil	Swell Expansion index test	ASTM D 4546-96 ASTM D 4829-95 or UBC 18-2
Shear strength for Slope stability	Unconfined compressive strength Unconsolidated undrained Consolidated undrained Direct shear Ring shear	ASTM D 2166-91 ASTM D 2850-95 ASTM D 4767-95 ASTM D 3080-90 ASTM D 4648-94
Erosion	Dispersive clay Erosion potential	ASTM D 4647-93 Day 1990b
Pavements and Deterioration	Pavements: CBR Pavements: R-value Sulfate	ASTM D 1883-94 ASTM D 2844-94 Chemical Analysis

Notes: * This specification is in the ASTM Standards Volume 04.03.
** These specifications are in the ASTM Standards Volume 04.09.
All other ASTM standards are in Volume 04.08

Laboratory testing usually begins when the subsurface exploration is complete. The first step in laboratory testing is to log in all of the materials (soil, rock or groundwater) recovered from the subsurface exploration. Then the geotechnical engineer or engineering geologist prepares a laboratory testing program, which basically consists of assigning specific laboratory tests for the soil specimens. The actual laboratory testing of the soil specimens is performed by experienced technicians, who are under the supervision of the geotechnical engineer. Because the soil samples can dry out or there could be changes in the soil structure with time, it is important to perform the laboratory tests as soon as possible.

Usually at the time of the laboratory testing, the geotechnical engineer or engineering geologist will have located the critical soil layers or subsurface conditions that will have the most impact on the design and construction of the project. The laboratory testing program should be oriented toward the testing of those critical soil layers or subsurface conditions. For many geotechnical projects, it is also important to determine the amount of ground surface movement due to construction of the project. In these cases, laboratory testing should model future expected conditions so that the amount of movement or stability of the ground can be analyzed. During the planning stages, specific types of laboratory tests may have been selected, but based on the results of the subsurface exploration, additional tests or a modification of the planned testing program may be required.

In all laboratory testing of rock and soil it is important to consider to what extent is the exploration sample and material representative of the insitu rock and soil mass. Representative sampling, test methods and engineering judgement are important factors.

For rock testing specifications, the most complete are those of the American Society for Testing and Materials (ASTM). These have been complemented by a series of suggested methods (Brown 1981) prepared by committees of the International Society for Rock Mechanics (ISRM).

In the case of rock laboratory tests, there are three main objectives in rock testing:

a) To provide basic information on the physical properties and mechanical reactions of the rock material.
b) To classify or characterize the rock material by providing an index which can be used to compare the particular rock with other rocks.
c) To provide information which can be used to design structures in the rock.

Some rock tests are sophisticated for the use to which they are put, and design in rock is often based more on field measurement and empiricism than on laboratory test data. Often there is increasing emphasis on the need for large numbers of quick insitu tests to give an indication of rock reactions rather than for detailed information from a particular test.

1. Laboratory Tests

    a)   Classification
- i) Density; moisture content; porosity; absorption
- ii) Uniaxial tensile and compressive strength and deformation characteristics
- iii) Anisotropy indices
- iv) Hardness; abrasiveness; attrition
- v) Permeability
- vi) Swelling and slake durability
- vii) Sonic velocity
- viii) Micro petrographic descriptions

    b)   Engineering Design
- i) Triaxial compressive strength and deformation characteristics
- ii) Direct shear tests
- iii) Time dependent and plastic flow characteristics

2. Field Observations and Tests

    a)   Characteristics
- i) Discontinuity orientation; spacing; roughness; geometry, etc.
- ii) Core recovery; RQD; fracture frequency; RMD
- iii) Insitu sonic velocity
- iv) Geophysical borehole logging

    b)   Engineering Design
- i) Plate and borehole deformability tests
- ii) Direct shear tests
- iii) Field permeability measurement
- iv) Insitu rock stress determination
- v) Post construction monitoring of rock movements
- vi) Insitu uniaxial, biaxial and triaxial compressive strength

3. Permeability

Permeability is defined as the ability of water to flow through a saturated porous media. A high permeability indicates that water flows rapidly through the void spaces, and vice versa. A measure of soil permeability is the coefficient of permeability, also known as the hydraulic conductivity *k*. The coefficient of permeability can be measured in the laboratory by using the constant head or falling head permeameter. The following table summarizes the coefficient of permeability versus soil type and drainage property.

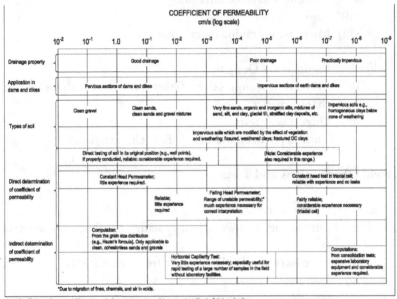

9.1 Coefficient of permeability versus drainage property, soil type, and method of determination
   (Developed by Casagrande, with minor additions by Holtz and Kovacs 1981)

In summarizing the importance of detailed laboratory testing results, it is relevant to consider what Tomlinson (1986) states:

> *"It is important to keep in mind that natural deposits are variable in composition and state of consolidation; therefore it is necessary to use considerable judgement based on common sense and practical experience in assessing test results and knowing where reliance can be placed on the data and when they should be discarded. It is dangerous to put blind faith in laboratory tests, especially when they are few in number. The test data should be studied in conjunction with the borehole records and the site observations, and any estimations of engineering design data obtained from them should be checked as far as possible with known conditions and past experience."*

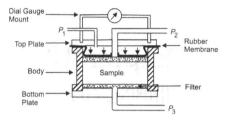

9.2 Soil Laboratory Testing Consolidation Cell

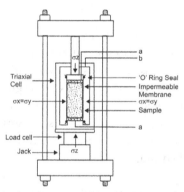

9.3 Soil Laboratory Testing Triaxial Cell

Chapter 10

HYDROGEOLOGY

The most observable part of the hydrologic cycle involves rainfall that runs downhill and accumulates into streams and rivers that combine and eventually reach sea level. Less obvious is the evaporation that simultaneously takes place, rises to form clouds, and returns to the ground surface as rain or snow or rejoins the hydrologic cycle. Even less obvious is water that infiltrates into the soil where its influences are manifest in springs and wells. These factors plus the influence of transpiration from vegetation and recycling by humans complete the hydrologic cycle.

The percentage of precipitation that infiltrates into the ground is close to 100% in cavernous limestone areas or desert sands and decreases to 10 to 20% on impervious materials such as in shale or granite. Much depends on the rate of rainfall and the moisture condition of the soil or rock. The measurement and prediction of rainfall and runoff are the realm of the surface water hydrologist, and the measurement and flow of groundwater is that of the groundwater hydrogeologist.

Runoff water is the main sculptor of landscape, and erosion patterns and the resistance to erosion are important clues to the compositions of rocks and soils.

Groundwater is the subsurface water that occurs beneath the water table in soils, rocks and geologic formations that are fully saturated. Groundwater also encompasses the near surface, unsaturated, soil moisture regime that plays an important role in the hydrologic cycle; and it includes the much deeper, saturated regimes that have an important influence on geologic processes.

Water Beneath the Surface of the Earth

- Groundwater exists in the spaces between grains of soil and fractures in bedrock.
- It represents the largest reservoir of fresh water available to humans (30 times more than lakes and rivers).
- It provides drinking water for 50% of the world's population.

Distribution

- Saturation is the maximum amount of water the air can hold at a given temperature and pressure.
- The water table is the upper level of a saturated zone of groundwater.
- A perched water table is a localized zone of saturation that is above the main water table. It is formed by an impermeable layer called an aquiclude or aquitard. Conversely, an area that allows groundwater to move easily is called an aquifer.
- The zone of aeration is the area above the water table where, soil, rock and sediment openings are unsaturated and filled with air.

Springs

- Springs are flows of groundwater emerging naturally at the surface.
- Most are spontaneous emissions of water caused when their surface intersects the water table.
- Hot springs are igneous dependent and contain water that is, on average, 6 degrees (C) warmer than the surrounding air.
- Geysers are formed when a fountain of hot air is periodically ejected into the air.
- Wells, either natural or artificial, are formed by an opening bored into the zone of saturation, that area where all the open spaces in rock or overburden soils are filled with water.
  - i) This, in turn, forms a cone of depression in the water table immediately surrounding the well.
  - ii) The difference in height between the bottom of this cone and the original water table is called a drawdown. .

- In subartesian wells, the water will rise above the top of the water bearing aquifer. In artesian wells, the water will rise above the ground surface.
- Wells can either be flowing (artesian) or nonflowing (subartesian).

Groundwater Depletion

- The world's water table has been dropping in some regions reportedly at a rate as fast as 1 metre per year.
- Even if this were reversed today, it could take 1,000 years to replenish the water table to its original level.
- This subsidence of the water table is causing damage to highways, roads, and bridges.
- It has irretrievably altered gradients, causing greater flooding along the flood plain and new flooding in areas where it was virtually unknown before.
- Groundwater in many parts of the world is becoming contaminated.
  - i) Purification depends on the type of substrate under the ground.
  - ii) Limestone, which is cavernous, does not promote purification. Sandstone, which is permeable, does promote purification.
  - iii) Since it takes many years to undo the damage, if it can be undone, the only solution often becomes to abandon a well which may be the life blood of an area.
- Groundwater plays a major role in the geological makeup of an area because it dissolves rock over time.
  - i) Groundwater is vital because it sustains streams and rivers when there is no rainfall.
  - ii) Subsurface erosion causes sinkholes and possibly unwanted caves to form. They can lead to ecological problems and/or structural damages.

Caverns

- There are thousands of caverns, naturally formed chambers, on Earth
- They form when acidic groundwater follows a weakness in rock, joints and bedding planes and hollows them out.
- The hanging and protruding rock formations in a cavern are stalactites, which are hollow and point down from the ceiling, and

stalagmites, which point up from the floor. A good mnemonic to remembering the difference is stalacties have a 'T' which has the end pointing down; stalagmites have an 'M' with the points up.

Karst Topography

- Consists of numerous sinkholes, i.e., depressions
- May form slowly over many years, or instantaneously
- Has only a few, short streams

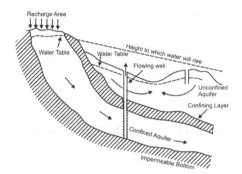

10.1 Schematic Cross Section
Confined and Unconfined Aquifers

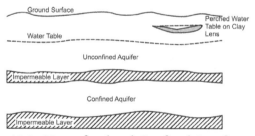

10.2a Unconfined and Confined Aquifers,
with a Perched Water Table in the Vadose Zone

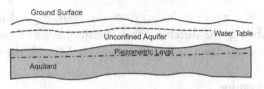

10.2b Leaky Aquifer

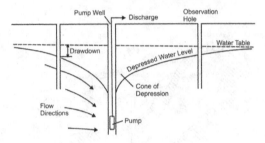

10.3 Pumping Test with Observation Wells

Chapter 11

GROUNDWATER

Groundwater Supply

For purposes of groundwater supply or groundwater control the geoengineering focus is on shallow and saturated groundwater flow zones.

Soils and pervious rocks contain a vast subsurface reservoir of water that has percolated downward to a zone of saturation. Groundwater is a major source for drinking and process water supplies. It helps to maintain lakes and streams during periods of low rainfall or drought. Water stored in soil pores is essential for survival of terrestrial plants and animals. The importance of groundwater may not be fully appreciated until a supply becomes depleted. This is a matter of considerable concern in areas where more water is being extracted as by irrigation than is being replenished by rainfall and snow melts. Climatic change as a result of uncontrolled development such as deforestation and industrialization can have a devastating influence on water supplies, as in the constantly expanding Sahara desert. North Africa is now arid, however once it was the breadbasket for Rome.

Water infiltrating into soil is drawn downward by gravity until it reaches a zone of saturation, where it starts to seep laterally to the nearest outlet that normally is a stream valley. The zone of saturation extends above the groundwater table because of capillary forces similar to those that draw water up into a fine tube. The level of the groundwater table therefore is measured in an open borehole or monitoring well that does not support capillary water.

The level, is also called the phreatic surface. It represents the elevation at which the pressure in the soil water is equal to atmospheric pressure.

The groundwater table is not an underground lake unless it happens to be in a cavern, and it is not flat like a table. Instead, the water table is higher under hilltops and declines to the level of nearby lakes or stream valleys, where groundwater may emerge as springs. Swimmers in lakes often will encounter zones of colder water caused by groundwater flowing into the lakes as springs.

Perched Groundwater

If downward-infiltration water encounters a relatively impermeable soil layer, the rate of infiltration may be slowed sufficiently to cause the water level to back up into the overlying soil, causing what is called a 'perched' groundwater table. Perched water tables sometimes contribute to landslides, and can be a problem in earth excavations.

Artesian Conditions

When a boring penetrates through an impermeable soil into a more permeable layers, water will rise up into the borehole and may even emerge or geyser out at the ground surface. This is an artesian condition and its occurrence and the boring depth should be carefully noted in borehole logs.

Tapping into an artesian aquifer can become out of control if water emerges rapidly enough to erode and open the borehole. The use of a drilling mud that is heavier than water can contain moderate artesian pressures, and if high artesian pressures are anticipated a special dense mud and/or casing should be used. Artesian conditions can be expected on floodplains that are capped with clay and are flanked by the flat terraces that gather water. An experienced driller must use hollow stem augers, casing, washboring and drilling muds in a boring to prevent a nuisance from becoming a disaster. Auger borings may even require the sacrifice of the auger if its withdrawal allows unimpeded erosive flow. Continuing escape of large amounts of artesian water will draw down the groundwater level and can dry up nearby wells as well as causing surface erosion and flooding.

A large hole, locally known as Jumbo, was described (Hardy and Spangler 2007) as the eighth wonder of the world in 1886 when artesian flow went out of control at Belle Plaine, Iowa, after the wrong size of wall casing was inserted into a groundwater supply well. Flow continued for over a year and finally was stopped by dumping in 40 railroad cars of rock, 25 tons of Portland cement, a diversity of scarp iron, plus unrecorded amounts of sand and clay.

Groundwater Level Fluctuations

The level of a water table changes depending on rainfall, snow melting and whether the ground is frozen or saturated when the snow melts. Curiously, rain can dry frozen soil by thawing and breaking through the frozen barrier that causes a perched water condition.

The management of infiltrating water is a major concern for geotechnical engineers as it influences such diverse factors as soil strength, reservoir levels, and soil conditions and erosion at construction sites. Water can reduce the bearing capacity of soils and is a key factor in landslides. Geotechnical engineers are concerned with seepage of water through earth dams, levees, and soil lining irrigation ditches, as well as flow into underdrains and wells. It is important that an engineer has a working knowledge of the principles governing the flow and retention of water in soil, and the effect of water on strength and stability of this material.

The groundwater supply considerations are:
- water as mankind's most vital and versatile resource
- origin of water
- formation of aquifer systems
- weather patterns and the hydrologic cycle
- occurrence and movement of groundwater
- groundwater chemistry
- groundwater resources
- groundwater exploration
- well hydraulics
- well drilling methods
- drilling fluids
- well screens and methods of sediment size analysis
- water well design
- installation and removal of well screens
- development of water wells
- collection and analysis of pumping test data
- water well pumps
- water quality protection for wells and nearby groundwater resources

- well and pump maintenance and rehabilitation
- groundwater law, water well specifications, and well contract problems
- groundwater monitoring technology
- alternative uses for well and well screens
- water treatment
- wise use of groundwater

Groundwater Control Needs

Groundwater control is often needed in civil and mining projects.

Any excavation (open or underground pit, tunnels, borings, caissons, tieback anchors, etc.) that penetrates below the groundwater level particularly in cohensionless soils will likely have seepage and soil erosion problems, which unless controlled will flood the excavation up to the original groundwater level.

The construction of deep excavations and foundations, underground openings and tunnels, underground services, etc. frequently require excavation into water bearing zones. The presence of water within soil affects its resistance to maintaining an open and stable excavation. In order to prevent caving or sloughing of the excavation slope or underground opening, dewatering and/ or stabilization as by grouting of the water bearing soil may be required to provide a dry and safe working area for work crews and equipment.

In cases where an excavation is undertaken near a water bearing stratum under excessive hydrostatic pressure, then heaving, blow outs and quick or flowing soil conditions may result in loss of the excavation base and/or walls. Depressuring of the saturated stratum by dewatering may be employed. Water bearing sand aquifers are problematic for open cut excavations (braced or sloped), tunnels, horizontal borings, and the like.

Dewatering systems must be carefully reviewed by the designers and specialists contractors for the local soil and groundwater conditions and to the depth of the excavation. A properly designed, installed and operated dewatering or depressuring system will:

a) intercept or control seepage entering into an excavation
b) increase the stability of excavated slopes and prevent slope failure

c) prevent quick, heaving or caving conditions from water bearing strata adjacent or beneath the bottom of an excavation by reducing excessive hydrostatic pressures

d) reduce lateral loads and earth pressures on temporary supports such as sheeting and bracing

e) improve excavation and handling characteristics of soil

When groundwater control methods are used, consideration should be given to the effect on local water wells. Improper handling of pump discharge may lead to environmental impacts, regulatory orders and complaints from local residents.

Well point and deep well systems must be adequately filtered to prevent soil erosion and pumped for a considerable length of time prior to excavation. They can lower groundwater levels in a wide area. A record of changing groundwater levels in the excavation area should be kept. This can be achieved by installing well points at key locations and observing the water levels regularly, that is, prior, during and after excavation. This information can forewarn of any deficiencies in the effectiveness of the groundwater control system and be helpful in the evaluation of local well interference complaints.

Groundwater Control Methods

Groundwater control may be carried out by two main methods; those that depend on natural groundwater flow and drainage processes, and those induced methods that keep the water from entering the excavation. A combination of the two may be used in some cases.

a) Collection Ditches and Sumps

This is usually for small excavations at or near water bearing strata where the soil or rock is not likely to slough or erode. The seepage is collected in ditches and removed from the excavation by sump pumps. This system is unlikely to be satisfactory where inflowing water causes quick or unstable soil conditions.

b) Sheeting and Open Pumping

Excavation into water bearing soils may be achieved by driving steel sheet piling to sufficient depth around the perimeter of the excavation, removing the earth and pumping out seepage as it enters the excavation.

Rapid lowering of the water level within the excavation as compared to the natural groundwater level outside of the excavation, can result in a build up of considerable hydrostatic pressure on the sheeting and on the bottom of the excavation. Sheeting must be adequately designed to withstand high earth pressures, braced and be driven to seepage cutoff depths. These can be determined from geotechnical borehole logs. Sheet piling may be used in conjunction with slurry trench cutoffs, cement or chemical grouts and cofferdams of various construction.

c)   Well Points

This system usually includes an array of 50 mm diameter screened wellpoints which are jetted in around the excavation at closely spaced centres of about 1.5 m. The wellpoints are attached to a common header pipe, which is connected to a suction pump. A wellpoint system is used effectively for temporary groundwater control in materials such as gravelly sands, silty sand and similar soils.

If the local groundwater level must be lowered more than 5 m (the maximum effective lift of suction pumps) below ground level, two or more stages may be needed.

Where deep excavation into relatively low permeability soils such as silty sand producing small amounts of water are necessary, wellpoints of a slightly larger diameter may be equipped with an aspirator or educator pump, with the pump intake near the bottom.

Vacuum and electro-osmosis methods may be used for dewatering silts and sandy silts where wells are ineffective in lowering the groundwater levels. The vacuum system is a wellpoint system, which is connected to a vacuum pump. By reducing the pressure in the water bearing strata water squeezes into the wellpoint and is removed. Soils that are stabilized by the vacuum method remain completely saturated with water and severe shocks can produce a collapse of loose soil structures resulting in quick and flowing soil conditions.

Electro-osmosis simply utilizes wellpoints as negative electrodes to attract water. Rods are driven into the soil to act as anodes. The wellpoints are connected to a header pipe and to a suction pump. The water flow is induced into the wellpoints by an electrical current applied to the rods. The total flow from a conventional wellpoint system is significantly increased when electro-osmosis is used.

d) Deep Wells

These are usually 150 to 450 mm diameter screened wells surrounded by a sand and gravel filter located at a relatively wide spacing around the perimeter of the excavation. Water is pumped using a submersible or turbine pump. Deep wells are used effectively where high yield, highly permeable cohesionless soil strata are encountered and where the depth of excavation below the local groundwater level is about 10 m. Deep wells or bleeder wells may be the only practical means of controlling uplift pressures where water bearing strata underlie impervious layers.

e) Compressed Air

This method of groundwater control is used exclusively in underground openings and tunnels. The tunnel is sealed off and compressed air is injected at a rate of about 10 kPa per metre of water level above the base of the excavation. This method is usually in conjunction with tunnel support structures.

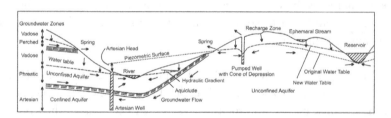

11.1 Aquifer Conditions Showing Varying Groundwater Flows

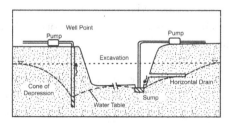

11.2 Groundwater Control System for Deep Sloped Excavation

Chapter 12
ROCK ENGINEERING

Rock engineering has been traced to the early days of civil engineering and mining by the Egyptian and Greek civilizations who quarried and transported large limestone blocks, constructed the pyramids in Egypt about 4700 years ago, crafted architectural columns, and built dams in Egypt and Iraq around 2900 B.C.

Rock mechanics is the study of the reaction of rocks to the forces imposed on them or in them by earthworks, mining and construction, often involving excavation and support.

Rock mechanics is the theoretical and applied science of the mechanical behaviour of rock and rock masses. It is that branch of mechanics concerned with the response of rock and rock masses to the force fields of their physical environment. Rocks are often layered, but more importantly they are fissured and jointed, and this means that rock masses may sometimes be controlled more in their reaction to forces by the discrete nature of the fissured mass than by the properties of the material. Rock mechanics should be considered as the study of rock deformation and fracture in both its intact massive form and as a discontinuous fractured form.

Both the practice of rock engineering and the applied science of rock mechanics are becoming more interconnected as man makes more demands on the Earth's resources and available space.

The field of engineering rock mechanics, as applied in mining engineering practice, is concerned with the application of the principles of geoengineering mechanics to the design of the rock structures generated by mining activity.

Rocks observed at the surface of the Earth today show deformation which ranges from various types of folds to different types of fractures. The deformations, which are mainly due to tectonic stresses, have been influenced by the lithology (Earth crust rocks) and the mechanical properties of the rocks in which they occur.

Rock is not made to design specifications. In rock engineering, where the rock itself is both the construction material and the structure, its properties have to be established from laboratory and field tests. Rock must be recognized as a discontinuous material, which can be expected to have different properties in different locations and directions. What is encountered is not a newly fabricated material, but one which has been subjected to severe mechanical, thermal and chemical actions over millions of years.

In order to predict how rock will behave as an engineering material, certain sets of properties have to be determined for:

- the intact rock;
- the fractures;
- the whole rock mass.

The relative importance of these different properties depends upon the particular engineering applications, but it is vitally important that the structural geology is fully appreciated, including strike, dip, stratification, folding, faulting, shearing, jointing, etc. as well as lithology (the rock type).

A standard item of field equipment for rock mapping is a device called a compass clinometer, which is used to measure the orientation of bedding, foliation, fault-planes and so on. For example, when the geologist is visiting isolated exposures of a sedimentary rock and wants to determine if the bedding in each one has the same orientation, otherwise the rock sequence has probably been folded.

A compass clinometer has a magnetic needle, functioning as in any ordinary magnetic compass, and a second, nonmagnetic, needle that is pivoted from one end so as to hang vertically under its own weight. The first measurement is to imagine a horizontal plane intersecting the geological surface such as a bedding plane whose orientation is to be determined. The two planes must intersect at a horizontal line and the angle between this line and north is defined as the strike of our geological surface. To measure the strike, the compass clinometer is held horizontally, with one edge pointing along the strike direction. The angle that this edge makes with the free-floating magnetic compass needle is the strike, and can be read off the dial. The second measurement is how steeply the geological surface is dipping, relative to horizontal. This is called the dip of the surface. To determine this, the compass clinometer is held with one edge running directly down the steepest gradient that can be found on the geological surface. The dip can then be read using the free-hanging needle.

In the chapter illustration, the strike is 330° and the dip is 60°. This would be recorded as 330/60 W or 330/60 WSW to be more accurate.

Within the geological context, and given the properties of the intact rock and the fractures, the other main design considerations are the stress state in the rock, hydrogeological conditions and the construction and mining procedures.

The term intact rock refers to rock which has no through fractures significantly reducing its tensile strength. Its properties are most important in rock excavation. It is usually characterized by density (unit mass), deformability (Young's modulus and Poisson's ratio), and strength (unconfined compressive strength, cohesion and angle of friction).

It is difficult to provide an accurate characterization of intact rock where its properties vary widely over the area of engineering interest. Moreover, even simple indicators of a property, e.g. Young's modulus, have to be qualified if the rock has different properties in different directions.

This leads to two quite different approaches to specifying intact rock properties:

- direct measuring of fundamental properties such as deformability;
- index testing as a comparative indication of intact rock quality.

The former is more relevant to theoretical analysis; the latter is easier and less costly to undertake. It is possible to conduct many index tests on relatively large quantities of rock, thus providing a better overall assessment of rock mass quality, rather than performing fundamental tests at specific locations.

All rock contains fractures. These are of varying types and occur on varying scales (from microfissures, through fissures, joints and bedding planes, to faults). The origin and interpretation of the structural geology of these fractures are of great assistance to engineers in indicating the mechanical structure and properties. For example, many sedimentary rocks are divided by major discontinuities into large blocks. Often the discontinuities are uniform in orientation and persistence, but the joints which fracture these blocks internally are frequently much less consistent in direction and extent.

The term discontinuity is used in rock engineering for all such types of fracture to indicate that the rock is not continuous, unlike the intact rock described above which is mechanically continuous. It is important to note that the nature, location and orientation of discontinuities significantly affect most of the rock properties (deformability, strength, permeability, etc.), and therefore the rock engineering applications.

Rock engineering projects include mining, quarrying and civil engineering needs as the creation of useable underground space. Rock slopes may be formed for a highway or during open cast mining. Tunnels and shafts have the purpose of access whether they are for a deep mine or for a mass transit system, but the design approaches can be very different. Similarly, mining and creating a permanent cavern both involve excavation and support, but they have different objectives: in mining only what has economic value is excavated, and collapse is prevented only to maintain access to the working face; whereas with a permanent excavation the underground space or cavern must not collapse, its shape or volume is the purpose of excavation, and the excavated material is wasted to spoil.

For caverns, new civil engineering applications are being developed, such as underground storage. The different methods of mining are areas to which the theory of rock mechanics has been applied, and there is minor precedent practice relating to the recent subjects of geothermal energy and radioactive waste disposal. In all of these applications, whether established or new, the principles of rock mechanics are important.

Foundations

Rock is usually an excellent foundation material, but near surface rock can be significantly fractured. It is always necessary to establish the competence of the rock to bear the required load at acceptable levels of deformation or settlement.

Rock Slopes

There are four basic mechanisms for rock slope failure: plane, wedge, direct toppling, and flexural toppling. The potential for failure in any of these modes can be easily identified using rock mechanics methods. The need and scope for a more detailed analysis can then be assessed.

Shafts and Tunnels

The stability of shafts and tunnels depends on rock structure, rock stress, groundwater flow, and construction technique. The fundamental principles of rock mechanics provide extremely useful guidelines for stability assessment.

Caverns and Underground Spaces

Rock joints have a major influence on the design and construction of large caverns. Methods to reinforce and support the rock are based on the principles of ground movement resulting from excavation for underground spaces.

Mining

There is a wide variety of geometries for open cut and underground mining, but in all cases the mining methods are designed to extract the mineral with minimum artificial support.

Geothermal Energy

In extracting geothermal energy, cold water is pumped down one borehole, to pass through fractures in a hot rock reservoir and exit from a second borehole. The optimal configuration for a production system depends on the interactions between rock joints, insitu stress, water flow, temperature and time.

Radioactive Waste Disposal

The aim is to isolate the waste so that unacceptable quantities of radionuclides do not return to the biosphere. Predicting the safety of a repository requires an understanding of all the factors listed above for geothermal energy, as well as others such as radionuclide sorption by rock fracture surfaces.

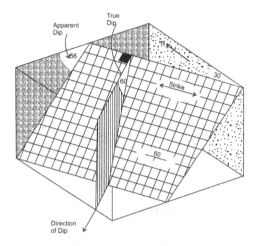

12.1 Strike and Dip
Orientation of cross hatched plane can be expressed as follows
Strike 330°, Dip 60°W; or Dip 60° towards 240°

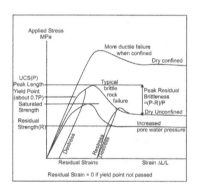

12.2 Stress Strain Relationships for Typical Rock

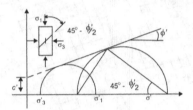

12.3a Mohr-Coulomb Failure Envelope for Soil

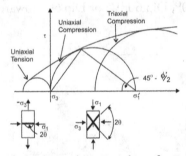

12.3b Mohr Failure Envelope for Rock
Rock is Weak in Tension

Chapter 13

SOIL MECHANICS

Soil mechanics is the study of the reaction of soils to the forces imposed on them or in them by earthworks, construction and mining.

1. Soil Formation

Soils in the engineering sense are either naturally occurring or man made. They are distinguished from rocks because the individual particles are not sufficiently bonded together.

Man Made Soils

The main types of man made ground are:

Waste Materials

These include the surplus and residues from construction processes such as excavation spoil and demolition rubble, from industrial processes such as ashes, slabs, mining spoil, quarry waste, industrial byproducts and from domestic waste in landfill sites. They can be detrimental to new works through being soluble, chemically reactive, contaminated, hazardous, toxic, polluting, combustible, gas generating, swelling, compressible, collapsible or degradable.

All random man made ground is non engineered and it should be treated as suspect because of the likelihood of extreme variability and compressibility.

These deposits have usually been randomly dumped and any structures placed on them will suffer differential settlements. There is also increasing concern about the health and environmental hazards posed by these materials.

Select Materials

Select materials have none or very few of the detrimental properties mentioned above.

They are used to form a range of earth structures such as highway embankments, earth fill dams, backfill around foundations and behind retaining walls. They are engineered fills that are spread in thin layers and are well compacted. This gives a high shear strength and low compressibility, to provide adequate stability and ensure that subsequent volume changes and settlements are small.

Contaminated and Polluted Soils

Due to past industrial activities many sites comprising naturally occurring soils have been contaminated (there are potential hazards) or polluted (there are recognizable hazards) by the careless or intentional introduction of chemical substances. These contaminants could comprise metals (arsenic, cadmium, chromium, copper, lead, mercury, nickel, zinc), organics (hydrocarbons, solvents, oils, tars, phenols, PCB, cyanide) or dusts, gases, acids, alkalis, sulphates, chlorides and many more compounds.

Several of these may cause harm to the health of people, animals or plants working in or occupying the site and some may cause degradation of building materials such as concrete, metals, plastics or timber buried in the ground.

Naturally Occurring Soils

The two groups of naturally occurring soils are those formed insitu and those transported to their present location. There are two very different types of soils formed insitu: weathered rocks and peat. Transported soils are moved by the principal agents of water, wind and ice although they can also be formed by volcanic activity and gravity.

Glacial Deposits

During the Pleistocene era which ended about 10,000 years ago the polar ice caps extended over a much greater area than at present with ice sheets up to several hundred metres thick and glaciers moving slowly over the earth's surface eroding the rocks, transporting rock debris and depositing soils of wide variety over northern Europe, the United States, Canada and Asia.

The deposits are generally referred to as glacial drift but can be separated into:

- soils deposited directly by ice;
- soils deposited by melt waters.

Wind Blown Soils

Wind action is most severe in dry areas where there is little moisture to hold the particles together and where there is little vegetation with no roots to bind the soil together. Wind-blown or Aeolian soils are mostly sands and occur near or originate from desert areas, coastlines and periglacial regions at the margins of previously glaciated areas.

Soil Particles

The nature of each individual particle in a soil is derived from the minerals it contains, its size and its shape. These are affected by the original rock from which the particle was eroded, the degree of abrasion and comminution during erosion and transportation and decomposition and disintegration due to chemical and mechanical weathering.

The mineralogy of a soil particle is determined by the original rock mineralogy and the degree of alteration or weathering. Particles can be classed as:

- Hard granular — grains of hard rock minerals especially silicates, from silt to boulder sizes.
- Soft granular — coral, shell, skeletal fragments, volcanic ash, crushed soft rocks, mining spoil, quarry waste, also from silt to boulder sizes.
- Plant residues — peat, vegetation, organic matter.
- Clay minerals — clay can have several meanings:

Clay soil — the soil behaves as 'clay' because of its cohesiveness and plasticity even though the clay mineral content may be small.

Clay size — most classification systems describe particles less than 200 µm; some soils less than 2 µm such as rock flour may not contain many clay minerals at all.

Clay minerals — these small crystalline substances with a distinctive sheet like structure producing plate shaped particles.

Soil Structure

The way in which individual particles arrange themselves in a soil is referred to as soil structure. This structure is sometimes referred to as a soil skeleton.

Granular Soils

Granular soils comprise coarse silts, sands, gravels, cobbles and boulders. Their soil structure will depend on:

- the size, shape and surface roughness of the individual particles
- the range of particle size (well graded or uniformly graded)
- the mode of deposition (sedimented, glacial)
- the stresses to which the soil has been subjected (increasing effective stresses with depth, whether the soil is normally consolidated or overconsolidated)
- the degree of cementation, presence of fines, organic matter, state of weathering, state of packing.

Cohesive Soils

Clay mineral particles are too small to be seen by the naked eye so their arrangements are referred to as microstructure or microfabric and our knowledge of particle structure comes largely from electron microscope studies.

There is a wide variety of soil particles, combinations of particles and structural arrangements produced by geological processes in nature. Each depositional environment has its own characteristics and post depositional processes can significantly modify the soils. When soils are affected by engineering structures or reworked during construction the particles and their structure may be irreversibly altered.

By knowing the undisturbed nature of soils and the effects of remoulding an assessment of the properties of soil such as shear strength, compressibility, consolidation, permeability, shrinkage, swelling, collapse, sensitivity, moisture content, unit weight, gradation, porosity, void ratio and relative density can be made.

When examining a soil and preparing a description it is important to convey sufficient information to those who have not seen the soil. It is often these persons who must make geological interpretations and assess the soil parameters and behaviour.

A systematic approach to soil description is based on the Unified Soil Classification System.

Classification tests include particle density, particle size distribution, bulk density, moisture content, liquid limit, plastic limit and shrinkage limit.

The classification tests are used to indicate the nature of the soil including parameters such as the liquidity index. They have a major contribution to the classification of soils according to the Casagrande plasticity chart.

2. Permeability and Seepage

Groundwater flow through soils is the most common cause of instability problems on construction and mining sites when excavating below the water table, and in earth structures retaining water.

The fundamental law of groundwater flow in saturated soils is Darcy's Law which related the quantity of water flowing through a cross-sectional area to the hydraulic gradient flow by the coefficient of permeability, k.

k is related to various soil properties, particularly the void sizes and shapes and the mass or macrostructure within a soil deposit.

Laboratory tests can be carried out to determine values of k but they may not represent the insitu macro-fabric effects. Field tests carried out in boreholes and larger scale pumping tests are important.

From the theory of seepage the construction of a flow net can provide values of seepage quantity and seepage force and an assessment of the stability of earth structures with respect to piping, boiling and heaving can be made.

Insitu values of k can be determined. The sketching method for a flow net may be used.

The factors governing the performance of soil filters are important.

3. Effective Stress and Pore Pressure

The principle of effective stress is fundamental to the understanding of soil mechanics. It equates the internal stresses within the mineral grain structure (the effective stress) and within the water (the pore water pressure) to the external stress (the total stress).

All soils during their geological formation have been subjected to a stress history, comprising deposition or loading, erosion or unloading and other environmental processes.

A normally consolidated soil is one that has undergone deposition only. An overconsolidated soil is one that has undergone unloading usually due to erosion, but other glaciation processes can cause overconsolidation.

The horizontal effective stress in the ground is not the same as the vertical effective stress, they are anisotropic. They are related by the coefficient of earth pressure at rest, Ko.

When a soil element is subjected to a change of stress it will undergo consolidation if loaded, or swelling if unloaded.

The change of pore pressure caused by a change of a total stress can be determined using the pore pressure parameters.

Above a water table there is a zone of full saturation where the surface tension in the pore water can sustain water in all of the voids. Above this level the soil becomes partially saturated where the finest capillaries can sustain water up to the capillary fringe.

The effective stresses are enhanced above the water table due to the negative pore pressures.

The effects of swelling and shrinkage during wet and dry seasons cause considerable damage to properties especially where they are close to tree roots. Soil dehydration and settlement have been observed in sensitive marine clays during drought conditions and climate changes.

Frost action on soils and the formation of ice lenses causes considerable heave problems during the cold period followed by weakening and settlements during the thaw period. Nevertheless ground freezing is used in a positive way to provide temporary support to excavations.

4. Contact Pressure and Stress Distribution

- The contact pressure beneath a loaded foundation area depends on the compressibility of the soil and the stiffness of the foundation.
- The stresses obtained by various stress distribution methods may vary considerably from those experienced insitu due to the simplifying assumptions made.
- Most methods give the stress at the corner of the loaded area so the principle of superposition must be used to determine stresses at other locations.
- The stresses within a stratum of finite thickness are greater than those within an infinitely thick deposit.

5. Compressibility and Consolidation

The compressibility of a soil is usually represented by a void ratio to effective stress relationship. From the preconsolidation pressure the soil can be distinguished as either normally consolidated and likely to produce large settlements, or overconsolidated with much smaller settlements.

The process of consolidation comprises the volume reduction with time of a fully saturated soil as water is squeezed out of the pore spaces under the application of a load. From the solution of the one-dimentional theory of consolidation the pore pressure within a soil layer and the settlement of the layer can be determined at any time after the application of the load.

The rate of settlement is determined by the insitu permeability which depends on the macro-fabric of the soil. Laboratory values of the coefficient of consolidation, c_v should be used with caution.

Two- and three-dimensional consolidation will increase the rate of settlement.

Construction techniques such as precompression by surcharging (preloading) and the installation of vertical subsurface drains are used to increase the rate of settlements.

6. Shear Strength

Stresses and strains in soils in a two-dimensional plane can be analysed using the Mohr Circle construction.

The shear strength of a soil varies with the amount of strain produced. After a yield stress has been reached plastic strains occur and the soil structure initially undergoes strain-hardening followed by strain-softening.

The Mohr-Coulomb relationship is the most commonly adopted failure criterion in soil mechanics. Stress paths are a useful way of demonstrating the changes of stress required to cause failure.

A soil can fail in an undrained or a drained manner depending on its permeability and the rate of applied loading.

Due to its high permeability, sand usually shears in a drained manner. The shear strength of a noncohesive sand or granular soil is dependent on the confining stresses, the initial density, the particle sizes, angularity, uniformity coefficient and particle crushing strength. Loose sands undergo strain-hardening during shearing and do not achieve a peak strength while dense sands display dilatancy and undergo strain-softening after reaching a peak strength.

Due to the low permeability of a clay soil, the normal rates of loading in construction works often produce the undrained condition so the undrained strength in terms of the total stresses is required. For long term stability of structures such as a slope the strength in terms of the effective stresses is required. The triaxal test method is commonly used to determine the undrained and drained strength of a clay soil. The shear box test is used to determine the drained strength of a sand. Reconstituted specimens are used since it is virtually impossible to obtain an undisturbed sample of sand.

The critical state theory involving the concepts of a boundary state surface, a critical state condition, the inter-relationships between mean stress, deviator stress and volume, the effects of drainage conditions, elastic and plastic straining, yielding and hardening provides a framework for basic soil behaviour.

The residual strength of a clay soil is obtained when large shear strains have occurred on a thin zone or plane of sliding. The particles have been rearranged to produce a strong preferred orientation in the direction of the slip surface and the strength is the lowest available and a fraction of peak strength.

7. Lateral Earth Pressures and Retaining Structures

Pore pressures in a soil are hydrostatic. Earth pressures are seated in the soil structure and are dependent on the effective stresses. The magnitude of the earth pressure is determined by the amount and type of movement. The minimum pressures are the active pressures, when the retained soil is allowed to expand and the maximum pressures are the passive pressures when the soil is compressed. In both cases the full shear strength of the soil is mobilized.

The Rankine Theory is frequently used to determine earth pressures and applied when the wall friction is zero, i.e. for a smooth wall. The Coulomb wedge theory is used for rough walls when friction is operative and for general cases of inclined walls and inclined backfill. Earth pressure coefficients are used to overcome the limitations of the Coulomb theory.

The effects of a cohesion intercept, the presence of a water table, surcharge loadings and compaction pressures are important.

There is a wide variety of retaining walls available for different circumstances. They tend to group into gravity walls that are constructed and then backfilled behind and embedded walls that are driven, bored or excavated into the ground before any excavation takes place.

The design of a gravity wall must prevent the occurrence of deep-seated slip failure, overturning or rotational failure, excessive bearing pressure beneath the toe, bearing capacity failure and base sliding.

An embedded wall may act as a cantilever, when the fixed-earth condition within its depth of embedment maintains its stability. When an anchor or prop below the top of the wall provides a supporting force the free-earth condition is assumed in the embedment depth. The ultimate limit state is determined for the moment equilibrium of the forces acting and limit state design codes recommend the use of partial factors on loads and strengths.

Due to the many factors outside of soil mechanics the loads on struts in a braced excavation are calculated using an empirical method.

The incorporation of reinforcement in an earth structure provides significant economies of soil materials and space occupied and has a wide range of applications. Stability and serviceability of a reinforced soil structure are assessed by considering the external factors such as bearing capacity and settlements, and internal factors such as the capacity of the reinforcement to withstand rupture or pull-out.

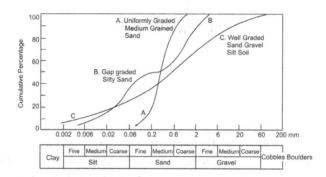

13.1 Soil Gradation Curve

Chapter 14

FOUNDATIONS

In civil engineering a foundation substructure links a superstructure with the earth on which it rests. The foundation is designed to transmit the weight of the structure, plus effects of live loads and wind loads, to the underlying material so that it is not stressed beyond its safe bearing capacity and tolerable settlements. If the supporting material is hard rock, design is simplified, but construction can become more costly because of the requirement for heavy ripping or blasting. If the depth to underlying rock is large, then the foundation rests on soil.

If the soil is not competent to support the anticipated load, the soil can be reinforced, or the load can be carried deeper to more competent soil or rock by means of piles or piers. The soil or rock support for a foundation is referred to as the foundation soil or foundation bed.

The most damaging type of shallow foundation results when the underlying soil shears and is displaced laterally, allowing the foundation to sink into the ground. This is a classic bearing capacity failure. Fortunately, it is rare. The concept is relatively straightforward, downward and sideways by shearing failure of the soil. The supporting ability of a soil depends in part on the depth of the foundation; the deeper the foundation, the greater the ability. Also eccentric loading and variations in shear strength inevitably cause the structure to tilt one way or the other. When the foundation softens slowly, but not catastrophically, then we refer to this as excessive total and/or differential settlement, which is more common. In this case, the soil consolidates and becomes stronger instead of shearing and becoming weaker.

An example of excessive settlement failure versus bearing capacity failure is the famous Leaning Tower in Italy, which started to lean soon after construction began, so the builders compensated for the lean by using thicker courses of stone on the low side. The correction temporarily relieved the off-center loading, but meanwhile soil beneath the low side had consolidated more than under the high side and was slightly stiffer, so tilting progresses in another area. The Tower stone courses zig-zagged upward. When it was completed there could be no more corrections, so tilting continued and then slowly tilted more as the load became more off center. If the soil became weaker through shearing instead of stronger through consolidation than the tower would have fallen over due to bearing capacity failure.

Bearing capacity failures can occur under road, railroad, or dam embankments if the rate of loading exceeds the capacity of the soil to drain away excess pore water pressure. In one case a heavily loaded train making a slow first run over a new railroad track slowed down and laid over on its side as the result of a bearing capacity failure.

A famous example of a bearing capacity failure from development of excess pore water pressure is the Tranconia grain elevator in Canada, which tilted about 45 degrees. After the silos were emptied they were righted by underpinning.

Bearing capacity failures are relatively rare except when an earthquake temporarily liquefies a sandy or silty foundation soil. During a 1964 earthquake in Niigata, Japan, a cluster of multi-storey apartment buildings sank and tipped to such a high angle that the occupants reached safety by crawling out of their window, and walking down the outside walls.

The first requirement in the design of a pile foundation is to choose the most appropriate and economical type of pile for the site conditions.

Besides the fundamental factors affecting a pile foundation it must be appreciated that the reliability of calculations based on analytical methods is often poor. Most building codes require that pile designs must be validated by static load tests carried out with the selected pile type and in the range of ground conditions anticipated.

Bored and driven piles in clay obtain most of their load capacity from shaft resistance whereas in sands driven piles achieve much higher base resistances. Bored piles in wet sand, particularly aquifers, are not recommended due to the adverse effects of drilling disturbances and running ground conditions. Preconstruction dewatering and drilling muds may be considered to stabilize wet sand conditions.

The limit state design approach can be carried out from pile load test results directly, or indirectly by validation of calculation methods based on ground test results, pile-driving formulae or wave equation analysis.

If negative skin friction is likely, but ignored, then a pile foundation will be subjected to settlements greater than anticipated. If it is allowed for then a more costly pile will result, with greater lengths required.

When piles are grouped together interaction effects produce load variations within the group and reduced efficiency of loading, especially for clays. The settlement of a pile group is greater than an individual pile and may bear no relationship to the load-settlement behaviour of a single pile.

Shallow Foundations

A shallow foundation generally derives its support from the soil or rock close to the lowest part of the structure, which it supports. The depth of the bearing area below the adjacent ground is usually about equal to or less than the width of the bearing area, and vertical loads on the sides of the foundation due to adhesion or friction may normally be neglected.

Shallow foundations include such common footing types as slabs, rafts, spread footings, strip footings, pads, mats and sills.

The design of a shallow foundation system normally requires that both bearing capacity and settlement be checked. Structural distress from soil settlement is often evidenced by cracking, distortion or openings. The drastic effects of a bearing capacity failure are rare except perhaps during construction where shallow temporary footings are frequently used with false work.

The general procedure is to define an ultimate bearing capacity, which is the anticipated failure load, and a safe bearing capacity, which is the ultimate capacity divided by a factor or safety. The usual factor of safety based on loads is 3 to 5, or lower subject to field testing.

Shallow foundations comprise strip, pad and raft foundations depending on the load applied and the shape required. Their design is based on the need to achieve stability requirements or an ultimate limit state and that they must satisfy settlement criteria and not be subjected to other ground movements referred to as serviceability limit states.

They are placed at shallow depths in the ground but not at ground level because of factors such as seasonal moisture variations, frost action, river erosion.

The stability condition is determined using bearing capacity theory applied to eccentric and inclined loading.

A method for the allowable bearing pressure of sand based on a limiting settlement is used.

The geotechnical design of footings is based on building codes which adopt limit state design. Partial factors are applied to characteristic values to give design values. The limit state is verified by ensuring that the design actions do not exceed the design resistance.

The determination of characteristic and representative values of soil parameters provides the geotechnical engineer with the greatest challenge.

The downward movement or settlement of a foundation is due to the stresses applied by a structure to the underlying soil. Settlements due to other factors must also be considered, such as seismic loads, groundwater lowering, soil dehydration, etc.

The total settlement of a foundation is the sum of the immediate, consolidation and secondary compression settlements. The immediate settlement will occur on application of the load so most should be completed by the end of construction. The consolidation settlement takes a longer time depending on the permeability of the soil and some of this will occur after the end of construction. The secondary compression occurs over a much longer period and can affect some structures for many years.

Immediate settlement is determined as the elastic undrained settlement at the corner of a flexible loaded area for both homogeneous and nonhomogeneous soil conditions. Corrections are available for the rigidity of the foundation, its depth and the possibility of local yielding of the soil. Using the principles of superposition and layering the settlement at any point near a loaded area and within the soil can be determined.

Consolidation settlement is determined using the compression index and swelling index method for normally consolidated and lightly overconsolidated clays. An accurate measure of the preconsolidation pressure is required and the insitu void ratio-log pressure curve should be constructed.

For more heavily overconsolidated clays the method using the coefficient of compressibility is often adopted with the Skempton-Bjerrum correction applied. The modulus of a soil increases with depth which will produce lower settlements. Some methods allows for nonhomogeneous soil conditions.

Settlements from secondary compression can be large for peat soils, normally consolidated highly plastic sensitive clays.

Settlement of foundations on sand are obtained from empirical correlations.

A criterion for the assessment of damage to a building can be based on the onset of visible cracking. By determining the settlement profile beneath the structure, the deflection ratio, Δ/L that could occur will determine whether the structure will incur damage or not.

An existing structure may be affected by movements of varying types. The effects of these movements include tilt, relative deflection, deflection ratio and angular strain.

Deep Foundations

A deep foundation is a foundation unit that provides support for a structure by transferring loads either by end-bearing to a soil or rock at considerable depth below the structure, or by adhesion or friction, or both, in the soil or rock in which it is placed.

Piles and caissons are the most common type of deep foundation.

Piles can be pre-manufactured or cast-in-place; they can be driven, jacked, jetted, screwed, bored or excavated. They can be of wood, concrete or steel. Bored piles of large diameter are frequently referred to as caissons.

The quality of a deep foundation system is highly dependent on construction technique, on equipment and on workmanship. Such parameters cannot be qualified nor taken into account in normal design procedures. Consequently, it is best practice to design deep foundation on the basis of insitu load tests on actual foundation units.

Instead of the use of piles, there are a number of techniques, which can be used to improve the strength and compressibility of subsoils, such as, vibro processes, preloading, grouting, chemical injection, freezing, and electro-osmosis.

Limit States Design

Limit States Design (LSD) is a design method used in structural engineering, and it is developing in foundation engineering. The method is a modernization and rationalization of engineering knowledge established before the adoption of LSD. In addition to the concept of a limit state, LSD includes the application of statistics to determine the level of safety required in the design process.

Limit States Design requires the structural foundation to satisfy two principal criteria; the ultimate limit state (ULS) providing sufficient strength and no collapse, and (SLS) having satisfactory performance and no occupant discomfort from settlement and cracking. A limit state is a set of performance criteria, such as adequate foundation soil strength and tolerable foundation soil settlement, that must be met when the foundation is subjected to loads.

Limit States Design has often replaced the previous concept of permissible or allowable stress design in many forms of civil engineering, but slowly in geotechnical and transportation engineering. It is often considered inappropriate to discuss safety factors when working with LSD. Australia, Canada, China, France, Indonesia and New Zealand utilize limit state design in their building codes. LSD is more adopted outside of the United States.

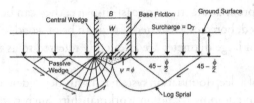

14.1 Bearing Capacity Failure Zones (Terzaghi)

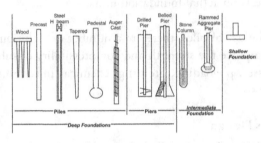

14.2 Deep, Intermediate and Shallow Foundations Types

Chapter 15

SLOPE STABILITY AND LANDSLIDES

The stability analysis of slopes plays a very important role in civil engineering and mining. Although level ground may be common sloping ground can be wide and deep where the land has been eroded by streams and rivers. If the slope is too high or too steep, or if the soil is too weak, the slope will find ways to correct the deficiency. A common corrective measure is a landslide event.

In terms of lives and property lost, landslides are among the most tragic and devastating events that involve soils. Landslides occur naturally as streams cut valleys deeper and wider, and as lakes and oceans erode their shores. Landslides can be initiated by shaking from an earthquake, or by the indiscretions of man.

Stability analysis is used in the construction of transportation facilities such as highways, airports, and canals; the development of natural resources such as surface mining, waste disposal, and earthfill and rockfill dams; as well as many other human activities involving building construction and excavations. Failures of slopes in these applications are caused by movements within the man made fill, in the natural slope, or a combination of both. These ground movements are studied from two different points of view. The geologists consider the ground movements as a natural process and study the cause of their origin, their pathways, and the resulting landforms. The geotechnical engineers investigate the safety of construction based on the principles of soil mechanics and develop methods for a reliable assessment of the stability of slopes, as well as the controlling and corrective measures needed. The best approach to slope stability studies is the combination of both these disciplines. The quantitative determination of the stability of slopes by the methods of soil mechanics must be based on a knowledge of the geological structure of the area, the detailed composition and orientation of strata, and the geomorphological history of the land surface. On the other hand, geologists may obtain a clearer representation of the origin and character of the ground movement process by checking their considerations against the result of geoengineering analyses based on soil mechanics. For example, it is known that one of the most favorable settings for landslides is the presence of permeable or soluble strata overlying or interbedded with relatively impervious strata.

Slope failures and landslides involve a variety of processes and contributing factors that have several different classifications, for example, they can be divided according to the form of failures, the type of materials moved, the age, or the stage of development.

An earlier reference on landslides or slope failures is a special report published by the Transportation Research Board (Schuster and Krizek, 1978). According to this report, the form of slope movements is divided into five main groups: falls, topples, slides, spreads, and flows. A sixth group, complex slope movements, includes a combination of two or more of the above five types. The type of materials is divided into two classes: rock and soil. Soil is further divided into debris and earth.

Recognizing the form of slope movements is important because they determine the method of stability analysis and the remedial measures to be employed. They are described as follows:

In falls, a mass of any size is detached from a steep slope of cliff, along a surface on which little or no shear displacement takes place, and descends mostly through the air by free fall, leaping, bounding, or rolling. Movements are very rapid and may or may not be preceded by minor movements leading to progressive separation of the mass from its source.

In topples, one or more units of mass rotate forward about some pivot point, below or low in the unit, under the action of gravity and forces exerted by an adjacent unit or by fluids in cracks. In fact, it is tilting without collapse.

In slides, the movement consists of shear strain and displacement along one or several surfaces that are visible or may reasonably be inferred, or within a relatively narrow zone. The movement may be progressive; that is, shear failure may not initially occur simultaneously over what eventually becomes a failure surface, but rather it may propagate from an area of local failure. This displaced mass may slide beyond the original failure surface onto what had been the original ground surface, which then becomes a surface of separation. Slides are divided into rotational slides and translational slides. This distinction is important because it affects the methods of analysis and control.

Classification of Slope Movements

Type of Movement			TYPE OF MATERIAL		
			Bedrock	Engineering Soils	
				Predominantly Coarse	Predominantly Fine
Falls			Rock fall	Debris fall	Earth fall
Topples			Rock topple	Debris topple	Earth topple
Slides	Rotational	Few units	Rock slump	Debris slump	Earth slump
	Translational		Rock block Slide	Debris block slide	Earth block Slide
		Many units	Rock slide	Debris slide	Earth Slide
Lateral spreads			Rock spreads	Debris spreads	Earth spreads
Flows			Rock flow (deep creep)	Debris flow (soil creep)	Earth flow (soil creep)
Complex			Combination of two or more principal types of movement		

In spreads, the dominant mode of movement is lateral extension accommodated by shear or tensile fractures. Movements may involve fracturing and extension of coherent material, either bedrock or soil, owing to liquefaction or plastic flow of underlying material. The coherent upper units may subside, translate, rotate, or disintegrate, or they may liquefy and flow. The mechanism of failure can involve elements not only of rotation and translation but also of flows; hence some lateral spreading failure may be regarded as complex.

Many examples of slope movement cannot be classed as falls, topples, slides, or spreads. In unconsolidated materials, these generally take the form of fairly obvious flows, either fast or slow, wet or dry. In bedrock, the movements are extremely slow and distributed among many closely spaced, noninterconnected fractures that result in folding, bending, or bulging.

Slope movements are divided into contemporary, dormant, and fossil movement stages. Contemporary movements are generally active and often recognizable by their configuration, because the surface forms produced by the mass movements are expressive and not affected by surface water and erosion. Dormant movements are usually covered by vegetation or disturbed by erosion so that the signs of their last movements are not easily observed. The causes of their origin remain and the movement may reoccur. Fossil movements generally developed in the Pleistocene or earlier periods, under different morphological and climatic conditions, and cannot repeat themselves as present slope movements.

According to stage, slope movements can be divided into initial, advanced, and exhausted movements. At the initial stage, the first sign of the disturbance of equilibrium appear and cracks in the upper part of the slope develop. In the advanced stage, the loosened mass is mobilized and slides downslope. In the exhausted stage, the accumulation of slide mass creates temporary equilibrium conditions.

Landslides may be classified according to their causes: (1) landslides arising from exceptional causes such as earthquake, exceptional precipitation, severe flooding, accelerated erosion from wave action, and liquefaction; (2) ordinary landslides, or landslides resulting from known or usual causes which can be explained by traditional theories; and (3) landslides which occur without any apparent causes.

The stability of slopes is a complex geoengineering problem which challenges any theoretical analysis.

Landslides occur when the downslope component of soil or rock weight exceeds the maximum resistance to sliding along a particular surface. Landslides can be shallow or deep. They differ from erosion by wind or running water because the main driving force is gravity acting on the weight of the soil or rock.

The best time to prevent a landslide is before it starts, because shearing and remolding of soil in the slip zone causes it to lose a substantial part of its shearing strength. After sliding starts it may accelerate, depending on geometric and geologic factors and how much strength has been lost. Sliding continues until the shearing resistance improves or the ground reaches a configuration that compensates for the loss of soil strength. Sensitive soils move farther and more rapidly then those that are less sensitive, sometimes becoming devastating mud flows that move so fast that they are inescapable.

Often for a landslide to stop it must move to a flatter level. This takes place gradually, as the sliding soil moves along a concave surface that is almost vertical at the crest and nearly horizontal at the toe. Passive resistance of soil building up at the bottom stops the slide. If that soil is removed, either naturally or as a result of man activities, the slide continues.

One of the worst things that can be done to encourage a landslide to continue seems to be the most intuitive, which is to put the soil back where it belongs. Often it takes two or even three attempts before owners are willing to admit defeat. Placing the soil back not only removes passive resistance, it puts weight where it will be most problematic for renewed sliding.

Man made slopes obviously must be designed to be safe from landslides. An active landslide automatically selects the most critical slip surface, but in a stable slope that surface is not known. Because of the infinite number of possible slip geometries the problem is statically indeterminate, so a slip surface either is assumed or is arrived at by computational trial and error.

Active landslides are more readily analyzed because the slip surface can be located by inspection and by borehole drilling. The analysis can determine the causes or if more then one causal factor is involved, prorate responsibility to the several contributors. Analysis is required to compare the effectiveness of different repair methods. Insurance policies most often have a waiver in fine print that excludes damages from ground movements, placing them in the category with nuclear wars and volcanic eruption.

Practically all the stability analyses of slopes are based on the concept of limit plastic equilibrium. First, a failure surface is assumed. A state of limiting equilibrium is said to exist when the shear stress along the failure surface is expressed as

$$\tau = \frac{s}{F}$$

In which τ is the shear stress, s is the shear strength, and F is the factor of safety. According to the Mohr-Coulomb theory, the shear strength can be expressed as

$$s = c + \sigma n \tan \Phi$$

in which c is the cohesion, σn is the normal stress, and Φ is the angle of internal friction. Both c and Φ are known properties of the soil. Once the factor of safety is known the shear stress along the failure surface can be determined.

In most methods of limit plastic equilibrium, only the concept of statics is applied. Unfortunately, except in the most simple cases, most problems in slope stability are statically indeterminate. As a result, some simplifying assumptions must be made in order to determine a unique factor of safety. Due to the differences in assumptions, a variety of methods, which result in different factors of safety, have been developed, from the very simple wedge method (Seed and Sultan, 1967) to the sophisticated computerized finite-element method (Wang, Sun, and Ropchan, 1972). Between these two extremes are the methods by Fellenius (1936), by Bishop method of slices (1955), by Janbu (1954, 1973), by Morgenstern and Price (1965), and by Spenser (1967).

In the stability analysis of slopes, many design factors cannot be determined with certainty. Therefore, a degree of risk should be assessed in an adopted design. The factor of safety fulfills this requirement. The factor should take into account not only the uncertainties in design parameters but also the consequences of failure. Where the consequences of failure are slight, a greater risk of failure or a lower factor of safety may be acceptable.

The potential seriousness of failure is related to many factors other than the size of project. A low dam located above or close to inhabited buildings can pose a greater danger than a high dam in a remote location. Often, the most potentially dangerous types of failure involve soils that undergo a sudden release of energy without much warning. This is true for soils subjected to liquefaction and that have a low ratio between the residual and peak strength.

The factors of safety suggested for mining operations in Canada, USA and Britain varies between 1.2 to 1.5.

For earth slopes composed of intact homogeneous soils, when the strength parameters have been chosen on the basis of good laboratory tests and a careful estimate of pore pressure has been made, a safety factor of at least 1.5 is commonly employed (Lambe and Whitman, 1969). With fissured clays and for nonhomogeneous soils, larger uncertainties will generally exist and more caution is necessary. A FS value of 2.0 and greater is used for high risk dams and slopes.

The emergency measures for slope failure and landslides can require immediate action to save properties from complete destruction:

a) If a structure is in imminent danger, local authorities must be informed and endangered structures evacuated.
b) Utility companies must be informed so that they can disconnect services or install flexible loops in the connections.
c) Surface water ingress must be stopped on and above the slide area.
d) Transverse tension cracks in the soil are plugged with soil to prevent entry of surface runoff water.
e) Surface runoff water is diverted away from the slide area with trenches, etc.
f) Backward tilting of the individual segments ponds water that infiltrates into the soil, so channels should be cut to drain the water.
g) As a temporary emergency measure a slope may be given a plastic cover that must be securely staked down to prevent blowing.
h) Other corrective measures include slope reduction, removal of surcharge weight, surface drainage, improvements, subsurface drainage, vegetation, buttress or retaining walls, pile systems for shallow slides, soil reinforcement, geosynthetics, anchor systems, hardening of soils by chemical, electro-osmosis and thermal treatment.

Mass movements or landslides on slopes can take several forms depending on the geological and hydrogeological conditions and the types of soil present. For some of these modes there is a method of analysis available and their degree of stability can be assessed but those that cannot be analyzed should not be ignored.

The stability of natural slopes and cuts depends on whether the slope has already failed at some time in the past. A first time slide will mobilize the peak strength initially, reducing to the critical state strength. The residual strength is to be expected where there is a pre-existing slip surface.

The stability of artificial earth structures, such as embankments and earth dams, is related to the failure of the slopes within the structure and failure within the foundation soils beneath. Placing and compacting fill materials means that slope failures are likely to be first time slides with the critical state strength mobilized. Foundation failures depend on the natural geological and groundwater conditions beneath and if soil with pre-existing slip surfaces is left in place then re-activation of slipping is likely to occur.

The slopes of embankments tend to fail in the long term after pore pressure redistribution has achieved a steady-state condition. Foundation failure beneath an embankment is most likely near the end of construction when pore pressures generated in the underlying soils are at their highest. Both the upstream and downstream faces of an earth dam can be subject to failure due to slope instability. Several other factors such as erosion, piping, hydraulic fracture are equally important.

Three methods of slope stability analyses are the plane translational mode, circular arc analysis using the method of slices and, wedge analysis. The degree of stability is represented by the factor of safety.

The effect of pore pressure on the stability of a slope is very significant.

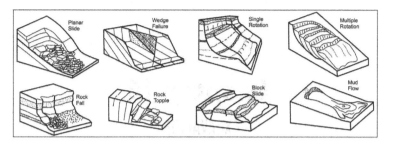

15.1 Types of Slope Failure and Landslides

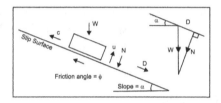

15.2 Forces Acting on Sliding Mass in Slope Failure Analysis

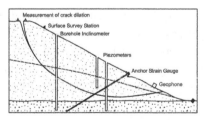

15.3 Slope Instability and Landslide Monitoring

Chapter 16

EXCAVATIONS

There are three main types of excavations used in construction and mining projects. Excavations may be (a) open cut sloped excavations where space permits, or (b) braced excavations using internal struts and rakers or tieback anchors where limited space is available, or (c) tunnels for shallow or deep underground openings.

Sloped Open Cut Excavations

The depth and slope of an open cut excavation and groundwater conditions control the overall stability and movements of open excavations. Factors that control the stability of the excavation for different material types are as follows:

a) *Rock.* For rock, stability is controlled by depth and slopes of excavation, particular joint patterns, insitu stresses, and ground water conditions.
b) *Granular soils.* For granular soils, stability usually involves side slopes, but does not extend significantly below the bottom of the excavation, provided that seepage forces are controlled.
c) *Cohesive soils.* For cohesive soils, stability typically involves side slopes but may also include the materials well below the bottom of the excavation. Instability of the bottom of the excavation, often referred to as *bottom heave*, is affected by soil type and strength, depth of cut, side slope and/or berm geometry, groundwater conditions, and construction procedures.

Slope stability analyses may be used to evaluate the stability of open cut sloped excavations in soils where the behaviour of such soils can be reasonably determined by field investigations, laboratory testing, and engineering analysis. In certain geologic formations stability is controlled by construction procedures, side effects during and after excavation, and inherent geologic planes of weaknesses.

Braced Excavations

Foundation works may require a relatively deep excavation with vertical sides. The sides may be supported by soldier piles with timber sheeting, sheet pile walls or diaphragm walls: these structures can be braced by means of horizontal or inclined struts or by tiebacks. In additional to the design of the supporting structure, consideration must be given to the ground movements which will occur around the excavation, especially if the excavation is close to existing structures. The following movements should be considered:

a)   settlement of the ground surface adjacent to the excavation,
b)   lateral movement of the vertical supports,
c)   heave of the base of the excavation.

To a large extent the above movements are interdependent because they are a result of strains in the soil mass due to stress relief when excavation takes place. The magnitude and distribution of the ground movements depend on the type of soil, the dimensions of the excavations, details of the construction procedure and the standard of workmanship. Ground movements should be monitored during excavation so that advance warning of excessive movement of possible instability can be obtained.

Assuming comparable construction techniques and workmanship, the magnitude of settlement adjacent to an excavation is likely to be relatively small in dense cohesionless soils but can be excessive in soft plastic clays provided that soil caving and groundwater seepage are controlled in the drilling methods.

Settlement can be reduced by adopting construction procedures which decrease lateral movement and base heave. For a given type of soil, therefore, settlement can be kept to a minimum by installation of struts or tiebacks as soon as possible and before excavation proceeds significantly below the point of support. Care should also be taken to ensure that no voids are left between the supporting structure and the soil. In cohesionless soils it is vital that groundwater flow is controlled; otherwise erratic settlement and sinkholes may be caused by a loss of soil into the drill holes and excavation.

The magnitude and distribution of lateral movements depends to a large extent on the mode of deformation of the supporting structure (e.g. whether the structure is allowed to deflect as a cantilever or whether it is braced near the surface with the maximum deflection taking place at greater depth). Lateral movement thus depends on the spacing and timing of installation of the struts or tiebacks. As in the case of settlement, excessive movements can occur if excavation is allowed to proceed too far before the first strut or tieback is installed. The other main factor is the type of soil. Under comparable conditions, lateral movements in soft to medium clays are substantially greater than those in dense cohesionless soils.

Most problems concerning braced excavations are the result of excessive ground movements and the control of such movements should be considered at the beginning of the design process. The design of the support system should be based on the requirements of movement control i.e. a serviceability limit state. There are three general approaches to the estimation of ground movements, namely (i) empirical correlations based on insitu measurements, (ii) the use of analytical procedures such as the finite element method and (iii) semi-empirical procedures which combine insitu observations with an analytical framework.

Base heave is generally a problem only in weak cohesive soils. The soil outside the excavation acts as a surcharge with respect to that below the base of the excavation, and therefore upward deformation, and in extreme cases shear failure, will occur. Short term heave will be mainly elastic, unless the factor of safety against base failure is low, but additional heave will occur due to swelling if the base remains unloaded for any length of time. In heavily overconsolidated clays, heave can be associated with the relief of the high lateral stresses existing in the clay prior to excavation.

Ground Anchors

A ground anchor normally consists of a high tensile steel cable or bar, called the tendon, one end of which is held securely in the soil by a mass of cement grout or grouted soil: the other end of the tendon is anchored against a bearing plate on the structural unit to be supported. The main application of the ground anchors is in the construction of tiebacks for diaphragm or sheet pile walls. Other applications are in the anchoring of structures subjected to overturning, sliding or buoyancy, in the provision of reaction for insitu load tests and in preloading to reduce settlement. Ground anchors can be constructed in sands (including gravelly sands and silty sands) and stiff clays, and they can be used in situations where either temporary or permanent support is required. Drilling anchors in wet sand aquifer conditions requires special care.

The grouted length of tendon, through which force is transmitted to the surrounding soil, is called the fixed anchor length. The length of tendon between the fixed anchor and the bearing plate is called the free anchor length: no force is transmitted to the soil over this length. For temporary anchors the tendon is normally greased and covered with plastic tape over the free anchor length. This allows for free movement of the tendon and gives protection against corrosion. For permanent anchors the tendon is normally greased and sheathed with polythene under factory conditions: on site the tendon is stripped and degreased over what will be the fixed anchor length.

The ultimate load which can be carried by an anchor depends on the soil resistance (mostly skin friction) mobilized adjacent to the fixed anchor length. This assumes that there will be no prior failure at the grout tendon interface or of the tendon itself. Anchors are usually prestressed in order to reduce the lateral displacement required to mobilize soil resistance and to minimize ground movements in general. Each anchor is subjected to a test loading after installation, temporary anchors usually being tested to 1.2 times the working load and permanent anchors to 1.5 times the working load. Finally, prestressing of the anchors takes place. Creep displacements under constant load will occur in ground anchors.

A comprehensive ground investigation is essential in any location where ground anchors are to be employed. The soil profile must be determined accurately, any variations in the level and thickness being particularly important. In the case of sands the particle size distribution should be determined, in order that permeability and grout acceptability can be estimated. The density index of sands is also required to allow an estimate of γ' to be made. In the case of stiff clays the undrained shear strength should be determined.

Trench Safety

Despite strict regulations and laws concerning safe practices in trenching, fatalities and damages occur from trench walls caving in on unprotected workers and equipment, respectively. The most common protection for workers in shallow trenches is a steel trench box that must be designed to withstand active soil plus impact pressures. The box is pulled along as workers install underground services in the open trenches, after which the trench is backfilled.

An engineer who observes dangerous violations of code is morally and professionally obligated to inform the contractor and, if necessary, the owner and the authorities – before it is too late.

There are few responsibilities in geotechnical engineering that carry the weight of human life so much as the design and construction of earth retaining walls. This is a serious business, and every design should be independently checked for subsurface parameters by an experienced geotechnical engineer and for structural stability by a specialist structural engineer.

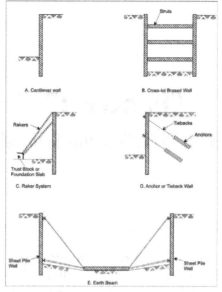

16.1 Common Types of Retaining Systems and Braced Excavations

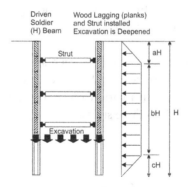

16.2 Pressure Diagram for Braced Excavations
(Terzaghi and Peck; 1967)

Chapter 17
LATERAL EARTH PRESSURES AND RETAINING WALLS

Earth retaining structures hold embankments, slopes and hillsides from sliding into structures and onto roadways. Bridge abutments are retaining walls that support the ends of bridges, and basement walls are retaining walls that serve a similar dual purpose, holding soil back while providing support for a building.

The design of retaining walls was first approached analytically by military engineers. An analysis made by Coulomb in the eighteenth century still is the basis for most computer programs.

The design of many subsurface structures such as retaining walls, anchored bulkheads, buried pipes, basement walls, braced excavations, thrust blocks, and others, require the determination of lateral earth pressures. Lateral pressures can be grouped into three states.

- Active K_a
- At rest K_o
- Passive K_p

The at-rest lateral pressure is the lateral pressure that exists in soil deposits that have not been subject to lateral yielding. The active and passive states of lateral pressure are limiting conditions and represent states of plastic equilibrium or at least partial plastic equilibrium. A state of plastic equilibrium exists when all parts of a soil mass are on the verge of failure. A state of active stress occurs when the soil deposit yields in such a manner that the deposit tends to move horizontally – for example, a retaining wall moving away from its backfill. A state of passive stress occurs when the movement is such that the soil tends to compress – for example, when a thrust block moves against the soil. The yield required to develop the passive state is much greater than for the active case.

Earth Pressure Coefficients

If a wall does not move, the intermediate stress state is *earth pressure at rest*. In this case the ratio of horizontal to vertical stress is designated by **Ko**.

A **Ko** condition occurs in nature during consolidation of sedimentary deposits, and occurs in a laboratory consolidation test because lateral strain is not allowed. Basement walls, bridge abutments, and walls that are keyed into adjacent buildings and are not allowed to move are designed to resist earth pressure at rest, which is higher than active pressure and lower than passive pressure.

Removal of overburden by erosion or excavation causes a proportionate reduction in vertical stress but does not significantly relief lateral stress. Some reduction of lateral stress will occur from elastic rebound, but for the most part it remains locked in. The inherited lateral stress in an overconsolidated soil therefore increases **Ko** so that it often exceeds 1.0, which would be impossible from application of vertical stress alone. Typical earth pressure coefficients for subsurface walls are **Ka** = 0.35, **Ko** = 0.5, and **Kp** = 3.0 subject to site specific assessments and excluding hydrostatic pressure.

Failure Mechanisms

Retaining walls must be designed to resist several different failure mechanisms that include:

- overturning;
- sliding along the base;
- building in the center area that may be preliminary to rupture;
- sinking and tilting as a result of eccentric loading and consolidation of the foundation soil;
- sinking caused by a foundation bearing capacity failure; and
- being part of landslide, referred to as global stability.

Retaining Wall Designs

The constantly increasing variety of retaining wall designs exhibits the ingenuity of engineers and builders.

Many retaining walls are gravity walls that are held in place by their own weight. These may be massive blocks of concrete or can involve a cantilever arrangement that uses weight of soil to bear down on a projecting heel of the wall to prevent tipping or sliding. Gravity walls normally are built from the bottom up.

Driven steel sheet piles have interlocking edges that create a continuous wall and have great versatility. They often are used for waterfront situations where dewatering is impossible, and for temporary walls, for example, to hold an excavation open and facilitate dewatering. This is a type of top-down wall. Tiebacks are installed as necessary to ensure stability.

A slurry trench uses pressure from a bentonite-water slurry to temporarily support the sides of a trench until the wall is built. Steel reinforcing cages are lowered into the slurry, which then is displaced from the bottom by pumping relatively dense fluid concrete through a canvas chute or tremie that reaches to the bottom of the trench and is raised as the trench is filled. After the concrete has set and gained strength, excavation proceeds on one side and tiebacks are installed to help support the wall. The walls sometimes are called diaphragm walls.

Yet another type of top-down walls is made by filling large vertical caisson borings with soil mixed with stabilizing chemical such as Portland cement, hydrated lime, or reactive fly ash. Adjacent borings overlap to create a continuous secant wall.

Tiebacks may be steel rods or cables with the ends grouted in place, or can be screwed-in soil anchors. Usually each tieback is pulled with a hydraulic jack, not only to test its anchoring capacity, but also to apply tension that helps to hold the wall in place, a process called post-tensioning.

A similar procedure called soil nailing essentially consists of the rod reinforcement without a wall, as steel rods hold the ground in place by friction. The exposed surface is protected from erosion by steel mesh and shotcrete, which is sprayed-on concrete.

In mechanically stabilized earth, or MSE, the wall functions as a decorative facing and is held in place by steel or plastic strips extending back into the soil, but is not post-tensioned.

Each type of wall uses lessons from the past, and can provide valuable insights that can be helpful in the future.

Influence of Surface Loads

Additional pressure is transferred to a retaining wall if a load is placed on top of the soil behind the wall, whether the load is from a building, a road, or a parking lot.

As a result, a load imposed on the surface of soil behind a wall ordinarily is not sufficient to overcome the shearing resistance of the soil, so the soil response is elastic. An elastic response tends to concentrate the additional pressure higher on the wall where it is more likely to cause tilting. It is for this reason that trucks should not be allowed to park close to the top of a retaining wall.

Quality Control During Construction

The internal friction of soil depends in part on the degree of compaction, so the compacted soil density and moisture content are part of the specification, particularly for a high wall. The usual procedure is to specify a required density and moisture content, which are measured with nuclear gauges. A more recent trend is to directly measure the soil internal friction with a rapid in-situ testing method, such as the Borehole Shear Test, as a quality control measure. This can eliminate the requirement of empirically correlating strength with density, and is particularly advantageous if the fill soil is variable.

ACTIVE, PASSIVE, AND AT REST EARTH PRESSURE

Internal Friction

Soil has characteristics of both a solid and a liquid, in that it exerts pressure against a vertical or inclined surface that increases approximately linearly with depth. However, unlike liquid, soil has an internal restraint from intergranular friction and cohesion. This is obvious because soil can be piled up, whereas a true liquid such as water flows out flat.

The ratio of lateral to vertical stress or pressure is designated by the coefficient *K*. One factor affecting soil *K* is whether the soil is pushing or is being pushed.

Active and Passive Earth Pressures

As previously described the upper limit of soil pressure is *passive earth pressure*, which develops as soil passively resists being pushed. The lower limit is called *active earth pressure* as soil acts to retain itself. *K* for the active case is designated *Ka*, and for the passive case, *Kp*. These are two of the most important parameters in geotechnical engineering, affecting not only pressures on retaining walls but also foundation bearing capacity, supporting strength of piles, and landslides. *Ka* is always less than 1.0, *Kp* is always greater than 1.0, and *K* for a liquid is 1.0.

Coulomb Analysis

While Rankine's and Coulomb's theories for soil pressure on a wall give identical answers, the Coulomb analysis includes a variable wall friction and is more suited for computer analyses. However, the analysis is a simplified version of what actually occurs in soil behind a retaining wall and can be on the unsafe side.

Coulomb reduced the retaining wall problem to a consideration of static equilibrium of a triangular single block sliding down a plane that extends down through the backfill to the heel of the wall. The triangular mass of soil sometimes is referred to as the sliding wedge. An analysis of the forces acting on the wedge at incipient failure reveals the amount of horizontal thrust that is necessary for the wall to remain in place. Coulomb's analysis is widely used even through experimental results indicate that the failure surface is curved instead of linear, and wall friction introduces arching action in the soil that is not considered in the analysis.

Design Consideration

The design of many structures requires the computation of lateral pressures using the Rankine and Coulomb principles. Lateral pressures are required to proportion the structure so that it will be both internally and externally stable. Conventional retaining walls, for example, must be externally stable against sliding and overturning, and soil pressure should not exceed the allowable bearing pressure of the soil under the toe of the wall. Moreover, the elements of the retaining wall must be adequately designed for movement and shear at all points. The design requirements of other retaining structures may be stated differently, but in any case the magnitude and distribution of lateral pressures will be required.

The magnitude and distribution of lateral pressures are functions of many variables and boundary conditions, including the yield of the structure, the type and properties of the backfill material, the friction at the soil structure interface, the presence of groundwater, the method of placement of the backfill material, and the foundation conditions for the structure.

Considerations in Designing the Backfill System

Backfill systems for retaining structures should be designed to minimize the lateral pressure that the structure must support. A good backfill material has two important attributes.

- High long term strength.
- Free draining.

In general, granular materials make the best type of backfill because they maintain an active state of stress indefinitely and are usually free draining. Clay soils, on the other hand, tend to creep and have a very low permeability. The tendency to creep causes clays to seek the at-rest case, which causes an increase in the lateral pressure with time. Thus, if a retaining structure has a clay backfill but was designed for active pressures, it will either fail structurally or deform to an extent that it becomes unusable.

Control of the water table in the backfill is also an important consideration. The total lateral force against a retaining wall for a fully submerged backfill will be two and one half to three times that for a dry backfill. Control of groundwater can be most easily accomplished with a free-draining backfill material. Some cases may also require an underdrain system in the backfill as well as control of surface drainage.

The extent of the backfill should be to the boundaries of the active wedge if the lateral pressures are to be computed on the basis of the properties of the backfill material.

The strength properties of soils usually improve with increased density therefore, it may seem desirable to compact the backfill material as much as possible, although this is generally not the case. Over compaction may increase lateral pressures. For non yielding walls excessive compaction can tend to increase the at-rest lateral pressure coefficient and, thus, the lateral pressure against the structure. Additionally, if heavy equipment is used to compact the backfill, the compaction equipment may induce loads on the structure that are much greater than the design loads. Therefore, it is important to consider carefully the backfill compaction requirements and to be specific in the construction specifications about the required compaction and the type of equipment that can be used.

It can be seen that in many situations involving earth pressures; the consideration of a simplified method is useful for the complicated interrelation between active and passive stresses and porewater pressures.

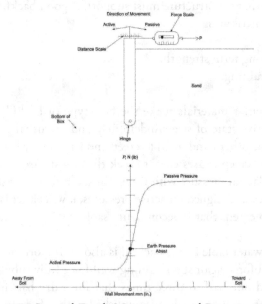

17.1 Lateral Earth Movements and Pressures

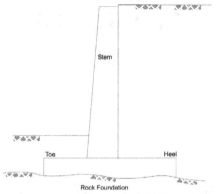

17.2 Cantilever Retaining Wall

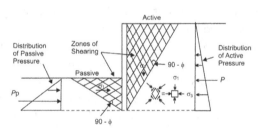

17.3 Idealized Rankine Stress Distribution with Zero Wall Friction

Chapter 18
TUNNELS

For many thousands of years, the main subterranean activities of man were cave dwelling, mining and tunneling. Mining was often carried out at considerable depths in ancient rocks, and the cavities excavated for this purpose were generally stable, with some form of structural lining being used as a means of preventing local falls of the roof or sides of the tunnel.

Tunneling, was often carried out at shallow depths in younger geological formations, and was used for water supply, drainage or military fortifications. The lining was needed to maintain the integrity of the cavity for conveyance purposes, and was designed as a permanent structure, resisting local loads by using brick or masonry in the form of vaulted arches. The tunnels of Babylon, Rome, Athens and Egypt were built this way, and there were few design changes throughout ancient history. Even in the Middle Ages, the substructures of our cathedrals and castles made extensive use of the masonry vault.

Between the seventeenth and nineteenth centuries, there were notable advances in the techniques of tunneling, owing to the introduction of explosives for blasting, and of hydraulic and pneumatic drills for rapid excavation. These advances coincided with a sharp increase in the need for traffic tunnels (highway, railway, navigation and subway) and conveyance tunnels (water supply, drainage, sewage, hydroelectricity), particularly in Europe and North America. Tunnels were normally constructed without disturbing the surface, using what became known as classical methods, where temporary timber elements in a variety of configurations were employed to support the heavy linings during erection. Later, Brunel invented the shield method, a moving metal casing driven in advance, to support the surrounding earth or rock without the need for timbers.

Lining materials changed from brick and masonry to concrete, reinforced concrete, cast iron and steel, particularly in the construction of the highway and subway tunnels in the early part of this century. Many shallow-depth urban tunnels were constructed by the cut-and-cover method, particularly if the soil was poor and saturated by groundwater. There was a need for all these structures to withstand the long term degeneracy associated with a soil and water environment, uneven bedding and fluctuating loads. They were rigid in construction, inherently strong and robust. Modes of failure were similar to those exhibited by masonry arches – the formation of hinges at the springlines and crown of the arch or vault, or inward failure of the lower sides due to lateral pressure followed by upward collapse of the floor. Structural analysis supporting the design was mainly concerned with establishing levels of loading and predicting failure of the heavy cross-sections. Many designers still employed thrust line theory to check the stability of the lining in the way it was used in the seventeenth century to design domes and vaults.

The NATM (New Austrian Tunneling Method) was developed for tunneling through rock, but uses principles from soil mechanics. Traditionally it was assumed that for reasons of safety a tunnel lining should be installed immediately as a tunnel is being advanced, thereby substituting lining support for high stresses existing insitu in deeply buried rock. The NATM allows a relaxation of these stresses before the lining is installed. Stress measurements in the lining indicate a substantial reduction in the amount of support required, and therefore in cost. Relaxation of stress occurs by elastic rebound and mobilization of supporting friction along fractures and bedding planes. The method involves careful monitoring of the rock expansion in order to select a proper time for installation of the liner. Then the liner plates are assembled inside the tunnel and grouted to fill the space between the rock and the plates.

A modern day tunnel is an underground stable opening of relatively uniform cross-section and significant length.

Of all engineering structures, tunnel excavations are most vulnerable to small troublesome and large catastrophic failures. One reason is because the tunnel excavation allows for the sudden release of large confining pressures resulting in rock strains which are not easily predicted. Another reason is the unpredictability of subsurface conditions, such as the possibility of encountering groundwater which can flood the tunnel. Because of the problems with tunnel excavations, tunnel coring or boring machines (for soft ground and hard rock) that can provide some protection during excavation of the heading, have been developed. Other measures to help stabilize tunnel rock faces include anchoring systems (such as bolts, rods, or dowels) and shotcrete. Chemical grouts have also been used to fill in rock discontinuities and joints in order to stabilize the tunnel crown and walls.

The ground conditions encountered in tunneling may be categorized as follows:

Hard: For hard soil, the tunnel heading may be advanced without roof support.

Firm: For firm soil, the roof section of a tunnel can be left unsupported for several days without inducing a perceptible movement of the ground.

Raveling: For raveling soil, chunks or flakes of soil begin to drop out of the roof at some point during the ground movement period.

Slow Raveling: The time required to excavate 5 ft (1.5 m) of tunnel and install a rib set and lagging in a small tunnel is about 6 hours. Therefore, if the stand-up time of a raveling ground is more than 6 hours, then the soil would be classified as slow raveling.

Fast Raveling: In contrast to slow raveling ground, for fast-raveling ground, chunks or flakes of soil drop out of the tunnel roof within 6 hours of being exposed by the tunnel excavation.

Squeezing: For squeezing soil, the ground slowly advances into the tunnel without any signs of fracturing. The loss of ground caused by squeeze and the resulting settlement of the ground surface can be substantial.

Swelling: For swelling soil, the ground slowly advances into the tunnel partly or chiefly because of an increase in volume of the ground. The volume increase is in response to an increase in water content of the soil. In every other respect, swelling ground in a tunnel behaves like a stiff non squeezing, or slowly squeezing, non swelling clay.

Running: The removal of lateral support on any surface rising at an angle of more than 34° (to the horizontal) is immediately followed by a running movement of the soil particles. This movement does not stop until the slope of the movement soil becomes roughly equal to 34°. If running ground has a trace of cohesion, then the run is preceded by a brief period of progressive raveling.

Very Soft
Squeezing: For very soft squeezing soil, the ground advances rapidly into the tunnel in the form of a plastic flow.

Flowing: Soil supporting a tunnel excavation cannot be classified as flowing ground unless water flows or seeps through it toward the tunnel. For this reason, a flowing condition is encountered only in free air tunnel below the groundwater tunnel or under compressed air when the pressure is not high enough in the tunnel to stabilize the excavation. A second prerequisite for flowing is low cohesion of the soil. Therefore, conditions for flowing ground occur only in inorganic silt, fine silty sand, clean sand or gravel, or sand and gravel with some clay binder. Organic silt may behave either as a flowing or as a very soft, squeezing ground.

Intact Rock: Intact rock contains neither joints nor hairline cracks. Hence, if it is breaks, it breaks across sound rock. On account of the disruption of the rock due to blasting, spalls may drop off the roof several hours or days after blasting. This is known as a spalling condition. Hard, intact rock may also be susceptible to the popping condition (i.e., rock burst) involving the spontaneous and violent detachment of rock slabs from sides or roof.

Stratified
Rock: Stratified rock consists of individual strata with little or no resistance against separation along the boundaries between strata. The strata may or may not be weakened by transverse joints. In such rock, the spalling condition is quite common.

Moderately
Jointed Rock: Moderately jointed rock contains joints and hairline cracks, but the blocks between the joints are locally grown together or so intimately interlocked that vertical walls do not require lateral support. In rocks of this type, both the spalling and the popping condition may be encountered.

Blocky and
Seamy Rock: Blocky and seamy rock consists of chemically intact or almost intact rock fragments which are entirely separated from each other and imperfectly interlocked. In such rock, vertical wall support may be required.

Crushed But
Chemically
Intact Rock: Crushed but chemically intact rock has the character of a crusher run. If most or all of the fragments are as small as fine sand-size particles and no recementation has taken place, then the crushed rock below the groundwater table exhibits the properties of water-bearing sand.

Squeezing
Rock: Squeezing rock slowly advances into the tunnel without a perceptible volume increase. The prerequisite for squeeze is a high percentage of microscopic and submicroscopic particles of micaceous minerals or of clay minerals having a low swelling capacity.

Swelling
Rock: Swelling rock advances into the tunnel chiefly on account of expansion. The volume increase is in response to an increase in water content of the rock. The capacity to swell seems to be limited to those rocks which contain montmorillonite.

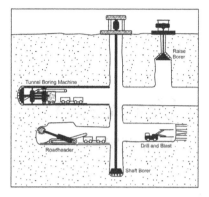

18.1 Three Methods of Tunnel Excavation
(i) drill and blast (ii) roadheader (iii) tunnel boring machine

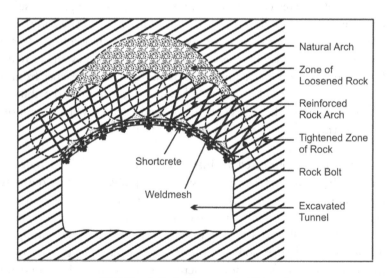

18.2 Rock Bolt Support in Tunnels

Chapter 19
EMBANKMENT DAMS

Dams

The geology of a water course valley, and the available supplies of suitable construction material, will influence the location of a dam site and the type of dam that is constructed. Beneath every dam is built a cut-off. This is a thin barrier that extends into the foundation and either prevents or reduces the leakage of reservoir water under the dam. The cut-off reaches from one abutment to the other and often extends some distance from the abutment into the side slopes of the valley. The depth and lateral extent of a cut-off is governed by the geology of the valley base and sides.

Every dam must be protected against sudden influxes of flood water into its reservoir, by an overflow structure such as a spillway, or other outlet that discharges downstream of the dam. To construct a dam it is necessary to divert the existing river and its flood waters either by diverting it through a channel to the side of the valley or diverting it into a tunnel that passes through the abutments and discharges downstream of the dam site. The geology of sites for appurtenance such as overflow structures and river diversion works must be considered.

The magnitude of ground accelerations coming from natural seismicity will influence the design of a dam and the auxiliary structures. Geological evidence of past earthquakes may be sought in regions where records are inadequate.

Types of Dams

There are three types of dam: (i) embankments made from sediment and rock (earthfill and rockfill are the terms used to describe suitable sedimentary material such as clay and sand, and rock blocks that can be placed to form a safe embankment), (ii) concrete dams and their forebears, the masonry dams, and (iii) composite dams, which are usually structures composed of more than one type of concrete dam but are occasionally composites that have concrete and embankment sections.

Embankment Dams

An embankment dam is often used where the foundation and abutment conditions are unsuitable for a concrete dam and where suitable materials for the embankment are present at or close to the site. An extensive ground investigation is essential, preliminary feasibility at first but becoming more detailed as design studies proceed, to determine foundation and abutment conditions and to identify suitable borrow areas. It is important to determine both the quantity and quality of available borrow materials. The insitu moisture content of fine-grained soils should be determined for comparison with the optimum moisture content for compaction of all engineered fills.

Embankment dams consist essentially of a core of impermeable material, such as rolled clay or concrete, supported by permeable shoulders of earth and rock fill. When a clay core is used it is normally flanked by filters or permeable material, such as sand, to protect the core from erosion by the seepage of reservoir water through the dam. Embankment dams, by virtue of the slopes required for their stability, have a broad base and impose lower stresses on the ground than concrete dams of similar height. Their fill is plastic and can accommodate deformations, such as those associated with settlement, more readily than rigid concrete dams. For this reason they can be built in areas where foundation rocks of high strength are not within easy reach of the surface; they are also the safest of all dam types against the risk of damage by earthquake. Their large volume requires copious supplies of suitable materials for earth and rock fill.

Size Categories of Dam

A dam is often grouped into one of the three following categories: (1) large dams, (2) small dams, and (3) landslide dams.

Large Dams

Large dams are more than 12 m (40 ft) in height and impound more than a one million m³ volume of water in the upstream reservoir.

Our Earth faces the hard truth, that less than 2.5% of our water is fresh, less than 33% of fresh water is fluid, less than 1.7% of fluid water runs in streams. Compounding this water shortage, one in five people worldwide lacks access to safe drinking water. Half the world lacks sanitation and millions die from waterborne disease. Farmers compete with cities for water. Municipalities drain aquifers that took centuries for natural processes to fill. Salt water pollutes groundwater inland from the sea. In a few decades, human civilizations may struggle to drink or bathe, and wars may develop over water as for oil.

It is reported that half of our world's rivers have been dammed at unprecedented rates, and at unprecedented scales of numerous large and high dams. Seasonally some rivers do not flow continuously, but stagnate in a string of reservoirs. In some years our world's largest rivers such as Africa's Nile, Asia's Yellow, America's Colorado, Australia's Murray may not reach the sea.

Countries build large dams over 12 m (40 ft) for sound reasons. Dams store, use and divert water for consumption, irrigation, cooling, transportation, construction, mills, power and recreation. Dams remove water from the Ganges, Amazon, Danube, Nile or Columbia to sustain cities along their banks. Dams are our oldest tool for hydraulic structures.

According to the International Commission on Large Dams (ICOLD), there are approximately 40,000 large dams on the world's rivers. Most large dams were built in the last 35 years. A further 1,600 are under construction in over 40 countries. Lobby groups for displaced rural communities and ecology organizations have disrupted dam building in the United States and India.

The links between large dam building and development are obvious. Two prerequisites for the development of a nation are energy and water. While dam building in developed countries has slowed in the last decade, major large dam constructions are underway in industrializing countries, like China's massive Three Gorges project and India's Narmada Valley Development project. Over half of all large dams (more than 22,000) are in China, while India has become the third largest dam constructor in the world, with over 3,000 large dams.

Dams produce power without contributing to the greenhouse effect. This is about 20 per cent of world electricity and seven per cent of all energy, according to ICOLD. Their primary purpose is water control. Reservoirs can provide drinking water, while smoothing out the cycles of flooding and drought brought about by monsoons. Large dams do this by storing excess water in reservoirs during the rainy season and releasing it in times of scarcity. The greatest use of dams is to supply irrigation water for agriculture, and one third of all food produced comes from irrigated land. Large dams with irrigation reservoirs are important for sustainable development, as about 80 per cent of our food production is expected to come from irrigated land by 2025.

Large dams are more than massive hydraulic structures with machines to generate electricity and store water. Large dams are concrete, rock and earth expressions of the technological age. Large dams are icons of economic development, scientific progress and engineering expertise to match internal combustion engines and nuclear power.

Small Dams

Small dams have been classified as those dams having a height of less than 12 m (40 ft) or those that impound a volume of water less than 1 million m^3 (1000 acre-ft) (Corns 1974). Sowers (1974) states that although failures of large dams are generally more spectacular, failures of small dams occur far more frequently. Reasons for the higher frequency of failures of small dams include: (1) lack of appropriate design, (2) owners believe that the consequences of failure of a small dam will be minimal, (3) small dam owners frequently have no previous experience with dams, and (4) smaller dams are often not maintained. Some of the common deficiencies with small dams are:

a) Subsurface or geological investigations are minimal or nonexistent.
b) No or little provision is made for future regular maintenance.
c) Slopes of the structures are constructed too steeply for routine maintenance.
d) Construction supervision varies from inadequate to nonexistent.
e) Hydrological design is deficient in that the top elevation of the dam is set an arbitrary distance above the pool with no flood considerations being made. These hydrological deficiencies are now amplified by development within the watershed.
f) Inadequate purchase of land to protect the investment in the project.

Landslide Dams

Landslide dams usually develop when there is a natural blockage of a valley by landslides, debris flow, or rock fall. Schuster (1986) states: *"Landslide dams have proved to be both interesting natural phenomena and significant hazards in many areas of the world. A few of these blockages attain heights and volumes that rival or exceed the world's largest man-made dams. Because landslide dams are natural phenomena and thus are not subject to engineering design (although engineering methods can be utilized to alter their geometries or add physical control measures), they are vulnerable to catastrophic failure by overtopping and breaching. Some of the world's largest and most catastrophic floods have occurred because of failure of these natural dams."*

According to Schuster and Costa (1986), most landslide dams are short lived. In their study of 63 landslide dams, 22% failed in less than one day after formation, and half failed within 10 days. Overtopping was by far the most frequent cause of landslide dam failure.

Common Causes of Earth Dam Failure

A dam failure has the potential to cause more damage and death than the failure of any other type of civil engineering structure. The worst type of failure is when the reservoir behind a large dam is full and the dam suddenly ruptures, which causes a massive flood wave to surge downstream. When this type of sudden dam failure occurs without warning, the toll can be especially high. For example, the sudden collapse is 1928 of the St. Francis Dam in California killed about 450 people; it was California's second most destructive disaster, exceeded in loss of life and property only by the San Francisco earthquake of 1906. As in many dam failures, most of the dam was washed away, and the exact cause of the failure is unknown. The consensus of forensic geologists and engineers is that the failure was caused by adverse geologic conditions which could have led to the disaster: (1) slipping of the rock beneath the easterly side of the dam along weak geologic plans; (2) slumping of rocks on the westerly side of the dam as a result of water saturations; or (3) seepage of water under pressure along a fault beneath the dam (Committee Report for the State 1928, Association of Engineering Geologists 1978, Schlager 1994).

Overtopping

According to Middlebrooks' (1953) comprehensive study of earth dams, the most common cause of catastrophic failure of an earth dam is overtopping, where water flows over the dam. This generally happens during heavy or record-breaking rainfall, which causes so much water to enter the reservoir that the spillway cannot handle the flow or the spillway becomes clogged. Once the earth dam is overtopped, the erosive action of the water can quickly cut through the shells and core of the dam.

Piping

The second most cause of earth dam failure is piping (Middlebrooks 1953). Piping is defined as the progressive erosion of the dam at areas of concentrated leakage. As water seeps through the earth dam, seepage forces are generated that exert a viscous drag force on the soil particles. If the forces resisting erosion are less than the seepage forces, the soil particles are washed away and the process of piping commences. The forces resisting erosion include the cohesion of the soil, interlocking of individual soil particles, confining pressure from the overlying soil, and the action of any filters.

There can be many different reasons for the development of piping in an earth dam. Some of the more common reasons are as follows (Sherard 1963):

- Poor construction control, which can result in inadequately compacted or pervious layers in the embankment.
- Inferior compaction adjacent to concrete outlet pipes or other structures.
- Poor compaction and bond between the embankment and the foundation or abutments.
- Leakage through cracks that developed when portions of the dam were subjected to tensile strains caused by differential settlement of the dam.
- Cracking in outlet pipes, which is often caused by foundation settlement, spreading of the base of the dam, or deterioration of the pipe itself.
- Leakage through the natural foundation soils under the dam.

Leakage of the natural soils under the dam can be a result of the natural variation of the foundation material. Any seepage erupting on the downstream side of the dam is likely to cause sand boils, which are circular mounds of soil deposited as the water exits the ground surface. Sand boils, if unobserved or unattended, can lead to complete failure by piping.

Clean uncemented sand and non plastic silt are probably the most susceptible to piping with clay being the most resistant. There can be exceptions such as dispersive clays. Dispersive clays are a particular type of soil in which the clay fraction erodes in the presence of water by a process of deflocculation. This occurs when the interparticle forces of repulsion exceed those of attraction so that the clay particles go into suspension and, if the water is flowing such as in a crack in an earth embankment, the detached particles are carried away and piping occurs.

According to Sherard (1972), one of the largest known areas of dispersive clays in the United States is north-central Mississippi. The cause of failure for several earth dams has been attributed to the piping of dispersive clays. The pinhole test is a laboratory test that an be used to classify the clay as being either higher dispersive, moderately dispersive, slightly dispersive, or non dispersive. The test method consists of evaluating the erodibility of clay soils by causing water to flow through a small hole punched through the specimen.

Slope Instability

Another common cause of dam failure is slope instability. There could be a gross slope failure of the upstream or downstream faces of the dam or sliding of the dam foundation. The development of these failures is similar to the development of gross slope failures and landslides and slope stability analyses can be used to design the upstream and downstream faces of the dam. Design analyses can be grouped into three general categories (Sherard 1963).

- Stability during construction, usually involving a failure through the natural ground underlying the dam. A total and/or effective stress slope stability analysis could be performed, depending on the type of soil beneath the dam and the rate of construction.
- Stability analyses of the downstream slope during reservoir operation. Usually an effective stress analysis is performed, assuming a full reservoir condition. A flow net can be used to estimate the groundwater level and porewater pressures within the earth dam.
- Stability analyses of the upstream slope after reservoir drawdown similar to item 1 above, the type of slope stability analysis (total stress versus effective stress analysis) would depend on the soil type and assumed rate of drawdown.

Besides the three most important design considerations (overtopping, piping, and slope instability), there can be many other analyses required for the design of earth dams. The checklist includes topography, geology, groundwater, weather, seismic vibrations, history of the slope changes, debris flow, creep, and the like for the design of dams. Depending on the size and location of the dam, the design and construction will have to meet various government regulations.

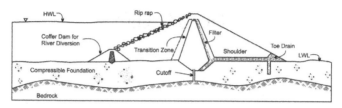

19.1 Cross Section of Embankment Dam

An impermeable core retains the reservoir and its underground extension of the cutoff prevents leakage that would otherwise occur beneath the dam. A slow flow of water occurs through these dams and is collected by a filter downstream of their core, which protects the core from erosion. This filter is connected to a basal drainage blanket through which water may readily leave the structure. The blanket also drains water seeping from the foundation strata downstream of the cutoff.

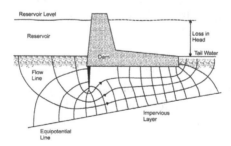

19.2 Flow Net Beneath Concrete Gravity Dam, with cutoff at the heel, seventeen equipotential drops and four flow channels.

Chapter 20

SOIL COMPACTION

Definition

Earthworks for excavations, cut and fill sections, dams, embankments, pavements, and the like, mostly have associated backfill and soil compaction requirements. The term engineered fill includes a specified native soil type, a select imported aggregate, optimum moisture content, and an adequate densification level.

A soil mass consists of solid particles surrounded by voids filled with water and/or air. Soil compaction is the process of forcing soil particles closer together. This is accomplished by eliminating, or at least reducing, the air voids in a given volume of soil.

Moisture content is the most important factor, because if the soil is saturated it cannot be compacted, since the voids are filled with water. Similarly, if the sample is dry, the film of water around the soil particle is missing, or is too thin to lubricate the particles when the compactive force is applied.

The objective in compacting soil is to improve its properties, in particular to:

a) increase its strength and bearing capacity;
b) reduce its compressibility and thereby minimize settlement;
c) decrease its ability to absorb water.

Compaction Testing

The usual method of determining the compaction achievement in the field is (a) to measure the inplace dry density, and (b) to relate it to a laboratory dry density, (c) for the same material. The laboratory dry density is established by the Proctor Test.

i)   Rubber Balloon Method

A hole is dug into the fill to be tested. The soil removed from the hole is weighed and its moisture content determined. The volume of the hole is then measured by pumping a water filled rubber balloon tightly into the hole from a graduated cylinder and thus the density is determined.

ii) Sand Cone Method

The only difference in this method from the balloon method is that the volume of the hole is determined by pouring sand of known unit weight into the hole.

iii) Nuclear Method

Nuclear moisture density gauges, although available since the early 1950's have been used extensively since the 1970's.

They contain a small radioactive source, which emits gamma rays to measure density, and neutrons to measure moisture.

In the field, the nuclear densometer gauge is placed on the surface of the soil to be tested and counts are taken on both density and moisture channels.

The field dry density is determined from the insitu moisture content and the bulk unit weight. The dry density is related to the Proctor value as with conventional testing equipment to obtain the percentage compaction.

Advantages and Disadvantages of the Various Testing Methods

Sand Cone

Advantages		Disadvantages	
a)	very accurate if meticulously executed and sand properly calibrated	a)	slow
b)	gives technician feeling of soil and has visual examination built-in	b)	subject to calculating errors
c)	permits stone correction	c)	moving and vibrating equipment has to be kept away
d)	Inexpensive	d)	requires numerous pieces of equipment
		e)	sensitive to weather influences
		f)	requires separate moisture content determination in laboratory

Balloon Apparatus

Advantages		Disadvantages	
a)	faster than sand cone	a)	requires skill and experience in pumping up balloon without rupturing it, or
b)	gives technician feeling of soil and has visual examination built-in	b)	danger that balloon will not be pumped up enough resulting in erroneous high densities
c)	permits stone correction	c)	should not be used in Granular A, crusher run and angular materials
d)	can be used while equipment working nearby	d)	requires separate moisture determination in laboratory
e)	relatively inexpensive		

Nuclear Gauge

Advantages		Disadvantages	
a)	fast (more results obtained)	a)	stone correction not possible
b)	allows moisture as well as density determination	b)	moisture channel not always reliable
c)	can be used while equipment working nearby	c)	does not give feeling of soil and does not allow visual examination without extra effort
d)	weather resistant	d)	requires knowledgeable operator
e)	nondestructive testing, if necessary	e)	electronic components subject to breakdown
f)	density profile can be taken with recent models	f)	expensive

Note: All the above field tests depend on an accurate soil identification to ensure that the laboratory Proctor density used is representative of the soil being tested.

Factors Influencing Compaction

The main factors influencing soil compaction are moisture content which is probably most important plus compactive effort, temperature, and grain size.

i) Moisture Content

At a low moisture content, the soil is stiff and difficult to compress. When water is added, the individual particles and/or lumps become coated with a film of water, which acts as a lubricant. The particles can slide along each other more easily. Up to a certain point, additional water replaces air and aids the compaction process. However, after a relatively high degree of saturation is reached, the water will occupy space, which would normally be filled by soil particles and the density decreases.

ii)   Compactive Effort

For most types of soil and compaction equipment, increasing the compactive effort will result in increased soil density.

Increased effort does not necessarily produce an equivalent increase in density as the maximum density for any soil is related to soil type, moisture content and the type of compaction equipment. Daily trial sections where the number of compactive passes versus field density is analyzed will generally provide important information to supplement the laboratory Proctor and field density tests.

iii)   Temperature

Although a minor factor in field compaction in warm climates, it can become a significant factor in cold regions with freezing temperatures affecting the moisture characteristics.

iv)   Grain Size

The grading characteristics can have a significant effect on the efficiency, as well as the effectiveness of compaction. If a material is poorly graded or uniformly graded (a material predominantly one size), it is difficult to compact. Temporary construction roads are sometimes constructed with a 300 mm layer of clear, crushed stone. It is difficult to compact such a layer. On the other hand, it is relatively easy to compact a Granular A material because of well graded particle sizes.

Validity of Insitu Density Measurement

The validity of insitu field density tests should be fully understood.

The maximum dry density is usually established in the laboratory by taking a representative sample of the soil and compacting it with a standard hammer into a mould at varying moisture contents. The maximum density obtained is termed the laboratory Proctor dry density. It is the density to which all field densities are related.

There are two laboratory Proctor density values, Standard Proctor dry density and Modified Proctor dry density. They are both obtained with similar procedures except that the modified value is determined by applying a greater standard compactive effort to put the material into the mould. Thus the Modified Proctor density is slightly higher than the Standard Proctor density. When specifying compaction values, therefore, it is important to ensure that you are relating to Standard or Modified Proctor values. Standard Proctor values are used in general earthworks construction, whereas Modified Proctor values are often used in airfield construction with high impact loads.

The Proctor dry density is not the maximum density that can be achieved in the field. Under ideal conditions, densities well in excess of the maximum Proctor value can frequently be achieved in the field.

It is important to understand what is inferred by 90% Standard Proctor dry density or 100% Standard Proctor dry density.

Most granular materials, when simply end-dumped from a truck and without any type of compaction, will achieve a density of approximately 83% to 85% of the laboratory Standard Proctor dry density. When compacted under ideal conditions, on the other hand, it is possible to achieve field densities of about 112% to 115% of the laboratory Standard Proctor dry density. Clays and silts, when end-dumped without compaction, are generally in the range of 78% to 82% Proctor density.

The relative compaction of soils in the field can vary from about a minimum of 80% Proctor density to 115%. Other Proctor densities outside of this range are erroneous.

The density measurements in the field and the Proctor tests in the laboratory are not an exact science. Compacted soils generally fall into five categories, namely, very poor compaction (less than 85%); poor compaction (between 85% to 90%); intermediate compaction (between 90% and 95%); good compaction (between 95% and 100%), and very good compaction (in excess of 100%). The rejection of soil compaction results when specified to be a minimum value of 95% SPMDD may not be justified because of a few density values of 94%. Because of variables, it is generally only possible, in practice, to estimate the relative densities to an accuracy of +/- 1%. Thus, calculated values of as low as 94% may be reasonable, particularly if the majority of values are in excess of 95%, and further field verification testing should be used to confirm this.

Compaction Equipment

There are two primary types of compaction equipment currently available namely static and vibratory equipment.

The static category includes all rollers such as smooth wheel, pneumatic tired, sheepsfoot, pack-all, etc. These static rollers are usually effective in cohesive soils such as clays and silts.

Vibratory compaction includes most of at the abovementioned rollers with a vibrating mechanism attached to the drum, as well as the vibratory plate tampers. These rollers and tampers should be used on cohesion less soils such as silt, sand and gravel.

Rammers are a third type of compaction equipment. Their use should be restricted to fill in confined areas where the other conventional compactors cannot be employed. Often the associated hydraulic forces can adversely affect adjacent structures and services.

Frequently Encountered Compaction Problems

i) Construction Problems

There are a number of problems which contractors encounter, including an incorrect type of compactor which is not suitable for the construction site and/or material; lifts of excessive thickness; underlying weak soil; and the moisture content of the backfill is too high or too low for achievement of the specified density.

The depth of loose layers to be compacted should not exceed 300 mm and with small plate tampers, it should be less than 150 mm to achieve a reasonable degree of soil compaction.

Sometime, it is impossible to achieve a high degree of compaction with imported granular fills since the underlying native soil is weak. This can be overcome by using an initial lift of clear stone which, in effect, firms up the weak soil to attain placement of additional lifts of granular fill.

As previously mentioned, moisture content control is critical if effective and efficient soil compaction is to be achieved. It may, for example, be desirable to protect stockpiled material with tarps to ensure that the moisture content is not increased during wet weather conditions. Conversely, in dry, hot periods, the addition of water is required. Once the material is in place and has been compacted to the specified level an increase or decrease in moisture conditions will not substantially reduce the relative density.

ii)   Technical Problems

In place densities are related to laboratory Proctor tests. If the soil or aggregate material changes, then a representative laboratory Proctor value must be obtained. Changing soil material is very common, in glacial areas, since the aggregate pit sources having been deposited during the last glacial melt era are heterogeneous in composition. Sometimes, these changes are subtle enough to escape the attention of a field inspector thus leading to inaccurate results.

After one or two metres of fill have been placed in layers the density results obtained at the exposed grade with either the balloon, sand cone or nuclear gauge do not represent the compaction for the total depth. In this case, tests pits have to be dug to ensure adequate coverage of insitu density at varying depths and all layers.

An earth structure can be formed from several different soil or fill types, each satisfying a different purpose. The feasibility, economics and efficiency of an earthmoving operation are dependent on the acceptability of the fill materials to be used.

The moisture condition test is usually specified to assess whether a soil is acceptable for use as a fill material.

In a temperate climate an earthworks project is heavily dependent on the weather conditions, due to the problem of the soil softening when wetted by rainfall and the adverse effects on earthmoving efficiency. Winter shut-downs are common.

Compaction of fill materials in the field, usually carried out by rollers, is effective in minimizing the air voids content and producing a high dry density provided that the appropriate type of roller is chosen for the soil type, maximum layer thickness is not exceeded, the minimum number of passes of the plant is adhered to, and the moisture content of the soil is within the specified limits.

Laboratory compaction tests provide the maximum dry density and optimum moisture content of a soil for a given compaction energy. They provide information on the suitable range of moisture contents for placement in the field and are used to provide control of field compaction in earthworks projects.

The properties of wet clay soils can be improved by the addition of lime and strong capping layers to support road pavements can be formed from lime stabilization. Cement-bound materials to form granular subbases and bases can be made from aggregates and granular soils with the addition of sufficient cement.

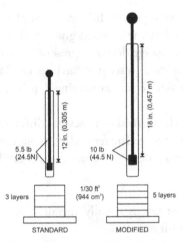

Proctor Maximum Dry Density
20.1 Laboratory Standard or Modified (SPMDD or MPMDD)

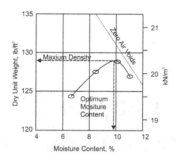

20.2 Soil Moisture Density Relationship Graph

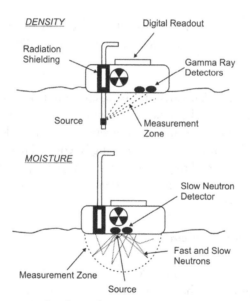

20.3 Principle of Nuclear Density Moisture Content Testing Device for Monitoring Field Compaction

Chapter 21

TRENCH EXCAVATION AND BACKFILL

Deep excavations may be cost effective sloped open cut excavations wherever lateral space permits, otherwise horizontally braced excavations may be needed.

In most instances, problems arise during open cut trench excavation work due to (i) unfavourable soil and groundwater conditions, (ii) excavation procedures, (iii) foundation and bedding requirements, or (iv) backfill compaction requirements. Well compacted pipe bedding is required to stabilize the native soil in the trench base and to provide adequate pipe support. Well compacted trench backfill is needed to support settlement sensitive roadways, walkways and nearby structures.

Trench Excavation Problems

Trench excavation can be undertaken, without significant problems in most types of soil, provided the water table is below the trench base. On the other hand, where excavation is attempted below a high water table, the considerations are very different. Before attempting to excavate, the location of the groundwater level and the soil type should be determined. It is very important to define these two parameters during a borehole investigation and test pit or trial excavation before the design and construction stages.

The groundwater level can be located by leaving the test pits open for at least one day. The soil type can be identified, by examining bulk samples from the test pits.

Having established both the groundwater level and soil type, the method of excavation in most soils can be assessed, excluding localized anomalies.

In consolidated clayey soils, the groundwater level for normal service pipe installation is not important. Since the soil is virtually impervious, even with a high groundwater level, excavation will be relatively straightforward, since negligible or no groundwater will enter the excavation area. Any water percolation into the trench is normally surface runoff and, it can be handled with normal construction drainage, that is, pumping from sumps and ditches. In soft clays, although groundwater control is generally not a serious factor, heaving of trench bases can occur with trenches in excess of 3 m depth. This problem can be overcome by sloping back the trench sides to flatter angles, and/or controlling underlying pressurized aquifers.

In sand, it is normally accepted that excavation to the groundwater level will be straightforward. In most instances, it should be possible to excavate approximately 300 mm below the groundwater level using perimeter ditches and filtered sump pumps. If the sand is very fine, then the temporary excavation may be extended to approximately 600 mm below the groundwater level. Under these circumstances, it is suggested that the installation be undertaken in as short a section as practical, i.e. short trench length excavated, the service pipe installed and then backfilled before the next section is excavated. Conversely, if the sand happens to be a uniform medium or medium/coarse sand, then a maximum depth of 300 mm below the groundwater level is suggested. Excavation below the above suggested limits without a positive method of controlling the groundwater level will lead to a serious problem because of increased groundwater flows. These positive methods include the use of wellpoints, or excavation within the confines of interlocking steel sheet piling driven to sufficient depth, or other groundwater controls.

Attempts to excavate trenches more than 300 mm below the groundwater level in a sandy soil is a problem. At many sites, the natural groundwater level is 1 to 2 m below ground surface. Thus excavation in wet sands to depths in excess of about 2 to 3 m require either wellpoints or sheeting. Wellpoints lower the groundwater level and therefore, the wellpoint tips should extend to a depth of at least 2 m below the proposed trench bottom. Sheeting, on the other hand, does not lower the water table. However, it prevents collapse of the trench sides and quick conditions or boiling of the base, when adequately designed and installed. A geotechnical engineer and a specialist dewatering contractor should be retained for this work.

The performance of excavations in wet silts is extremely difficult to predict. Several deep excavations have been undertaken without having to resort to sheeting. Wellpoints are not normally effective in dewatering fine silts since they have very low permeabilities. Attempts to excavate in fine silts stratified with coarse sand seams to coarse silts have met with serious trench excavation problems.

Trench Stabilization

The need for stabilization to firm up the base of a service trench prior to placing the pipe bedding is generally a function of the type of soil and the groundwater levels, and to some extent, the expertise of the contractor. In most areas, the soil in trench bases will tend to be easily disturbed, particularly in excavations slightly below the water table in sands without using well points, or in soft clays, or loose wet silts. The disturbance to the soil in trench bases can be increased with the continual passage of workmen and equipment, i.e. the trench base in wet silt becomes liverish.

In most cases where a base instability problem is anticipated, or occurs, slightly over excavating the trench to provide for a thin layer of well graded, angular, granular materials, firms up the disturbed soil and prevents further deterioration. Frequently success has been achieved with clear stone, since it is heavy and its angular fragments bind and cement together to produce a firm base. A 50 mm clear stone is suggested for this purpose in most instances. A 150 mm layer is adequate. It should be noted that 50 mm clear stone must be placed to a depth that can be driven into the poor soil trench base. If too deep a layer of clear stone is placed, the large voids that remain can fill with the migration of surrounding fines at a later date causing settlement. Well graded Granular A, B or C materials do not normally provide a good material to stabilize the trench base because of the large percentage of finer material which becomes unstable when wet.

Trench Backfill

Trench backfill is compacted in order to improve its properties and, in particular, to increase its strength and supporting properties, as well as reduce its compressibility and thereby minimize settlement. The compaction of trench backfill is essential in trenches beneath settlement sensitive structures such as buildings, roadways, parking areas, etc. It is not as critical where services traverse non settlement sensitive facilities such as parks and fields where the backfill does not have to support structures and settlement is not important, however ground surface settlement and depressions or swales can affect surface drainage.

Most pipes will normally be surrounded with a granular bedding material. Trenches are backfilled to the subgrade level with approved excavated native soil previously excavated from the trench to reduce differential heaving. The lower portion of the trench is backfilled with approved Granular C type sand and the upper 1.2 m with native soil material.

The effects of non compacted backfill in trenches should be considered. Studies have been performed on the amount of settlement associated with compacted fills. It can be assumed that any fill compacted to a minimum value of 95 percent Standard Proctor density will not result in settlement of any appreciable magnitude. Fill compacted to a lesser degree, say 90 percent, may result in settlement, as follows:

$$\frac{H}{A}(95\% - \text{actual compaction }\%)$$

where   H = total depth of loose fill
        A = 1 or 2 for various soil types

Thus, 2 m of fill, estimated to be compacted to approximately 90%, will settle approximately 50 to 100 mm. End dumped fill (compacted to 85%) will settle 100 to 200 mm, or more in some cases. This is an empirical relationship developed from experience and numerous case histories with soils in Ontario. It obviously produces an approximate value, although it is sufficient to provide a guide as to the effects of inadequate compaction of fills. It should be noted that most fills, in normal underground service pipe installation, are somewhat less than the total trench depth after deducting the granular bedding materials which have been placed over the pipe and the upper 600 mm of granular road base. Frequently, the actual depth of native material in a trench is limited to about 1.2 m. Settlements should be assessed for the poorly compacted fill zones only. In most cases, some compactive effort is applied to the fill either during placement or by construction traffic, thus the resulting settlements are almost negligible. Such settlements may be tolerated in granular roads, but not on paved roads. It is prudent to delay paving, particularly with surface coarse asphalt until at least one freezing and thawing season have passed.

Approved excavated native soil is generally placed back in the trench excavation. This continuity with the adjacent native soil is required to prevent severe differential frost heaving problems if an alternate granular material is substituted.

In many instances, it is unfortunately necessary to accept the lesser of the two evils, i.e. knowing that some slight settlement will occur in trench backfill with native backfill depending on the total depth, or no settlement but severe differential frost heaving movements if compacted granular fill is substituted. It has been found in the majority of cases that the settlement of native fills may be less than 50 mm whereas differential heaving can be at least 100 mm. In addition, once the settlement of the native fill has occurred generally within six months, minor additional movement will likely take place. On the other hand, differential frost heaving will occur repeatedly each year.

Backfilling with approved excavated native soil is a good practice. However, some caution should be exercised when the fill exceeds 2 or 3 m depth, because of potential settlement. In this instance, special efforts should be undertaken with attempts to compact the native soil, such as scarifying and discing to dry the soil to a moisture content suitable for compaction. Alternatively, the trench fill below the frost zone could be an imported Granular C, which is readily compacted.

Typical results of distress in pavements due to inadequately compacted trench backfill are often evident on the approach to bridge abutments and culvert pipes, etc., where the approach fill settles. However, the adjacent structures are rigid since they are constructed on foundations designed for minor settlement.

Peat and Bedrock

Finally, one or two special cases should be considered, such as, construction of underground service pipes in peat areas and bedrock. Most of the problems associated with pipe installation in rock are attributed to poor pipe bedding. It is important to note that rock and rock fill are notable poor insulators. Theoretical calculations, assuming an average number of freezing index days for a typical Canadian winter, indicate that frost can penetrate rock about 2 m. This can be compared with 1 m for clay and 0.5 m for fibrous peat. Pipe trenches, which cross both peat and bedrock require adequate transition zones to minimize differential settlement and cracking.

Construction in peat should be avoided, if possible, that is, the building site or pipe alignment should be relocated to avoid deep areas of peat. If this cannot be avoided, then whenever possible, the peat should be entirely excavated to a competent, non compressible soil and backfilled up to invert level with engineered fill. It is frequently practical to only excavate and remove peat to depths 4 to 5 m beyond which complete excavation and backfilling becomes economically marginal. However, peat has been excavated to depths of 8 m where localized deep pockets exist rather than run the risk of future settlement and cracking, etc.

Other muskeg treatments where deep deposits are encountered involve partial excavation and partial displacement of the organic by surcharging the soil with fill and inducing failure beneath the proposed roadway, thus displacing the organics from the construction alignment. This procedure is somewhat unreliable, since it cannot be carefully controlled and some muskeg will undoubtedly remain, resulting in future settlement and maintenance of the road.

Surcharging with blasting is possible, but is not normally recommended although some articles suggest this procedure in remote wilderness conditions.

Attempts have been made to float the pipe in peat. This approach needs very careful engineering consideration, in particular, the final condition in the trench loads placed on the peat underlying the sides of the pipe. For example, if the trench were to be backfilled with imported fill, then the actual load of the backfill would be much heavier (fill density of 1800 kg/m^3 or greater compared to peat density of 1020 kg/m^3). A pipe could be installed in peat, for example by compensating the loss of weight due to the void created by the pipe with some lightweight bedding. The trench may have to be backfilled with the peat. The lightweight bedding may be beneficial in tending to prevent the pipe floating. It is considered important to install the pipe and backfill in as short sections as possible, probably in the order of 10 to 20 m sections. On a trial basis, this scheme has proved to be practical and successful in some areas, however requires a great deal of trial and error and very close supervision at the onset of the project.

One spectacular failure of a sewer line occurred when the pipe was placed in compressible organic soil and was subsequently surcharged with 2.5 m of fill. The pipe settled immediately and was displaced laterally, as well. During construction and commissioning, one section of the pipe was displaced to such an extent that it was never recovered.

Pipes supported on piles driven through deep peat deposits into bedrock or a competent soil layer may be a positive solution, however lateral displacement must be considered.

21.1 Sloped Open Cut Trench Excavation

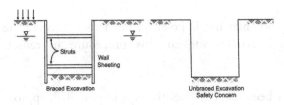

21.2 Braced and Unbraced Trench Excavation

Chapter 22
GEOSYNTHETICS

Geosynthetics are thin flexible, permeable sheets of synthetic material used to stabilize and improve the performance of soil associated with civil engineering works. Correctly designed and installed, geosynthetics have the ability to filter, drain, reinforce and separate soil. In many applications, geotextiles may be designed and selected to perform a combination of these functions. For example, when installed at the base of a granular fill embankment constructed over soft clay all four applications may be used.

Over a century ago, sheets of canvas were incorporated in earthfill to reduce lateral earth pressures exerted behind retaining walls. About fifty years ago, thick cotton fabric similar to denim, was used in the United States to stabilize dirt roads. Similarly, synthetic textiles were used by the Dutch as filters in the rapid repair of the North Sea dykes.

Since these early applications there have been continual advancements in the supply, design and application of geosynthetics in civil engineering structures.

A geosynthetic is defined as a planar product manufactured from polymeric material and typically placed in soil to form an integral part of a separation, filtration, drainage, reinforcement, moisture barrier and/or stabilization system. Common types of geosynthetics used during construction are geogrids, geotextiles, geomembranes, geonets, geocomposites, geopipes, geofoams, and geosynthetic clay liners (Rollings and Rollings, 1996).

Suitable test methods should be carried out to confirm the supplier's data on physical, mechanical, hydraulic and chemical properties, as follows:

a) Physical – fibre type (woven and nonwoven), length and width of roll, weight.
b) Mechanical – tensile, tearing and bursting strength, puncture resistance, elongation.
c) Hydraulic – water permeability, soil retention.
d) Chemical – resistance to light, ultra-violet, chemicals, micro-organisms, high and low temperatures.

1. Geogrids

A geogrid, contains relatively high-strength polymer grids consisting of longitudinal and transverse ribs connected at their intersections. Geogrids have a large and open structure and the openings (apertures) are usually 1.3 to 10 cm in length and/or width. Geogrids can be either biaxial or uniaxial depending on the size of the apertures and shape of the interconnecting ribs. Geogrids are principally used as follows:

i) Soil reinforcement. Geogrids are used for subgrade stabilization, slope reinforcement, erosion control (reinforcement), and mechanically stabilized earth-retaining walls, also used to strengthen the junction between the top of soft clays and overlying embankments.
ii) Asphalt overlays. Geogrids are used in asphalt overlays to reduce reflective cracking.

The common usage of geogrids is for soil reinforcement. Compacted soil is strong in compression but weak in tension. The geogrid is just the opposite, strong in tension but weak in compression. Thus, layers of compacted soil and geogrid tend to compliment each other and produce a soil mass having both high compressive and tensile strength. The open structure of geogrid allows the compacted soil to bond in the open geogrid spaces. Geogrids provide soil reinforcement by transferring local tensile stresses in the soil to the geogrid. Because geogrids are continuous, they also tend to transfer and redistribute stresses away from areas of high stress concentrations such as beneath a wheel load. Geogrids are used as soil reinforcement for mechanically stabilized earth retaining walls.

Similar to other geosynthetics, geogrids are transported to the site in 0.9 to 3.7 m wide rolls. It is generally not feasible to connect the ends of the geogrid, and it is typically overlapped at joints. Typical design methods for using geogrids are summarized by Koerner (1998).

Some of the limitations of geogrids are as follows:

a) Ultraviolet light. Even geogrids produced of carbon black (i.e. ultraviolet stabilized geogrids) can degrade when exposed to long-term ultraviolet light. It is important to protect the geogrid from sunlight and cover the geogrid with fill as soon as possible.

b) Nonuniform tensile strength. Geogrids often have different tensile strengths in different directions as a result of the manufacturing process. It is essential that the engineer always check the manufacturer's specification and determine the tensile strength in the main and minor directions.

c) Creep. Polymer material can be susceptible to creep. Thus, it is important to use all allowable tensile strength that does allow for creep of the geosynthetic. Tensile strengths are often determined by using ASTM test procedures, such as ASTM D 6637-01 ("Standard Test Method for Determining Tensile Properties of Geogrids by the Single or Multi-Rib Tensile Method,"2004) and ASTM D 5262-04 ("Standard Test Method for Evaluating the Unconfined Tension Creep Behaviour of Geosynthetics,"2004).

Many manufacturers will provide their recommended long term tensile strength for a specific type of geogrid. This recommended long term design tensile strength from the manufacturer is usually much less than the ultimate strength of the geogrid. The engineer should never apply an arbitrary factor of safety to the ultimate tensile strength, but rather obtain the recommended long term design tensile strength from the manufacturer.

2. Geotextiles

Geotextiles are the most widely used type of geosynthetic and they are often referred to as fabric. For example, common construction terminology for geotextiles include geofabric, filter fabric, construction fabric, synthetic fabric and road reinforcing fabric.

Geotextiles are usually categorized as being either woven or unwoven depending on the type of manufacturing process. Geotextiles are principally used as follows:

a) Soil reinforcement. Used for subgrade stabilization, slope reinforcement, and mechanically stabilized earth retaining walls; also used to strengthen the junction between the top of soft clays and overlying embankments.

b) Sediment control. Used as silt fences to trap sediment on site.

c) Erosion control. Installed along channels, under riprap, and used for shore and beach protection.

d) Asphalt overlays. Used in asphalt overlays to reduce reflective cracking.

e) Separation. Used between two dissimilar materials, such as an open graded base and a clay subgrade, in order to prevent fines contamination.

f) Filtration and drainage. Used in a place of a graded filter where the flow of water occurs across or perpendicular to the plane of the geotextile. For drainage applications, the water flows within the geotextile.

A common usage of geotextile is for filtration, i.e. flow of water through the geotextile. For filtration, the geotextile should be at least 10 times more permeable than the soil. In addition, the geotextile must always be placed between a less permeable soil and a more permeable open graded gravel material. An appropriate use of a geotextile would be to place it around the drainage pipe, because then it would have more permeable material on both sides of the geotextile and it would tend to restrict flow.

Geotextile to be used as filtration devices must have adequate hydraulic properties that allow the water to flow through them and they must also retain soil particles. Important hydraulic properties are as follows:

- Percent open area. Although geotextiles have been developed that limit the open area of filtration to 5% or less, it is best to have a larger open area to develop an adequate flow capacity.

- Permittivity or flowrate. Manufacturers typically provide the flow capacity of a geotextile in terms of its permittivity of flow rate. These hydraulic properties are often determined by using ASTM test procedures, such as ASTM D 4491-99, Standard Test Methods for Water Permeability of Geotextiles by Permittivity Soil Retention Capability, 2004.

- Apparent opening size. The apparent opening size (AOS), also known as the effective opening size (EOS), determines the soil retention capability. The AOS is often expressed in terms of opening size (mm) or equivalent sieve size. For example, AOS = 40-70 indicates openings equivalent to the No. 40 to 70 sieves. The test procedures in ASTM D 4751-99, Standard Test Method for Determining Apparent Opening Size of a Geotextile, can be used to determine the AOS. If the geotextile openings are larger than the largest soil particle diameter, then all of the soil particles will migrate through the geotextile and clog the drainage system. A common recommendation is that the required AOS be less than or equal to D_{85} (grain size corresponding to 85% passing).

Some of the limitations of geotextiles are as follows:

a) Ultraviolet light. Geotextiles that have no ultraviolet light protection can rapidly deteriorate. Manufacturers will often list the ultraviolet light resistance after 500 h of exposure in terms of the percentage of remaining tensile resistance based on the text procedures in ASTM D 4355-02, Standard Test Method for Deterioration of Geotextile by Exposure to Light, Moisture and Heat in a Xenon Arc Type Apparatus, 2004.

b) Sealing of the geotextile. When used for filtration, an impermeable soil layer can develop adjacent to the geotextile if it has too low an open area or too small an AOS.

c) Construction problems. Some of the more common problems related to construction with geotextiles are as follows (Richardson and Wyant, 1987):

i)   Fill placement or compaction techniques damage the geotextile.

ii) Installation loads are greater than design loads, leading to failure during construction.

iii) Construction environment leads to a significant reduction in assumed fabric properties, causing failure of the completed project.

iv) Field seaming or overlap of the geotextile fails to fully develop desired fabric mechanical properties.

v)   Instabilities during various construction phases may render a design inadequate even though the final product would have been stable.

3. Geomembrane

Common construction terminology for geomembranes includes liners, membranes, visqueen, plastic sheets, and impermeable sheets. Geomembranes are most often used as barriers to reduce water or vapor migration through soil. Another common usage for geomembranes is for the lining and capping systems in municipal landfills. For liners in municipal landfills the thickness of the geomembranes is usually at least 80 mil. The surface of the geomembranes can be textured in order to provide more frictional resistance between the soil and geomembrane surface.

Some of the limitations of geomembranes are as follows:

a) Puncture resistance. The geomembrane must be sufficiently thick so that it is not punctured during installation and subsequent usage. The puncture strength of a geomembrane can be determined by using the test procedures outlined in ASTM D 4833-00, Standard Test Method for Index Puncture Resistance of Geotextiles, Geomembrane, and Related Products, 2004.

b) Slide resistance. Slope failures have developed in municipal liners because of the smooth and low frictional resistance between the geomembrane and overlying or underlying soil. Textured geomembranes have been developed to increase the frictional resistance of the geomembrane surface.

c) Sealing of seams. A common cause of leakage through geomembranes is due to inadequate sealing of seams. The following are different methods commonly used to seal geomembrane seams (Rollings and Rollings, 1996):

    i) Extrusion welding: Suitable for all polyethylenes. A ribbon of molten polymer is extruded over the edge (filet weld) or between the geomembrane sheets (flat weld). This melts the adjacent surfaces that are then fused together upon cooling.

    ii) Thermal fusion: Suitable for thermoplastics. Adjacent surfaces are melted and then pressed together. Commercial equipment is available that uses a heated wedge or hot air to melt the materials. Also, ultrasonic energy can be used for melting rather than heat.

iii) Solvent-based systems: Suitable for materials that are compatible with the solvent. A solvent is used with pressure to join adjacent surfaces. Heating may be used to accelerate the curing. The solvent may contain some of the geomembrane polymer already dissolved in the solvent liquid or an adhesive to improve the seam quality.

v) Contact adhesive: Primarily suitable for thermosets. Solution is brushed onto surfaces to be joined, and pressure is applied to ensure good contact. Upon curing, the adhesive bonds surfaces together.

4. Geonets and Geocomposites

Geonets are three-dimensional netlike polymeric materials used for drainage, i.e. flow of water within geosynthetic. Geonets are usually used in conjunction with a geotextile and/or geomembrane, hence geonets are technically a geocomposite.

Depending on the particular project requirements, different types of geosynthetics can be combined together to form a geocomposite. For example, a geocomposite consisting of a geotextile and a geomembrane provides for a barrier that has increased tensile strength and resistance to punching and tearing.

5. Geosynthetic Clay Liners

Geosynthetic clay liners are frequently used as liners for municipal landfills. The geosynthetic clay liner typically consists of dry bentonite sandwiched between two geosynthetics. When moisture infiltrates the geosynthetic clay liner, the bentonite swells and creates a soil layer having a very low hydraulic conductivity, transforming it into an effective barrier to moisture migration.

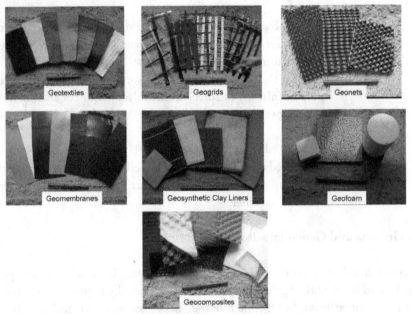

22.1 Typical Geosynthetic Materials
(Terrafix Geosynthetics Inc.)

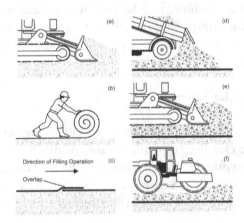

22.2 Six Stages of Typical Construction Techniques for Geosynthetics

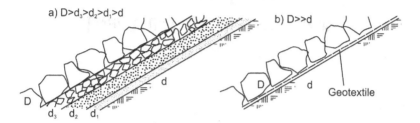

22.3 Filter System Alternatives Under Rip Rap Slope Protection
a. graded aggregate filter system
b. geotextile filter system
D = diameter of armour and d = diameter of filter or subsoil

Chapter 23

PAVEMENTS, CONCRETE AND ASPHALT

Historical Background

Although pavement design has evolved from art to science, empirical methods are important to the present day. Prior to the early 1920s, the thickness of pavement was based on experience. The same thickness was used for a section of highway even though widely different soils were encountered. As experience was gained, various methods were developed by different agencies for determining the thickness of pavement required.

Because the thickness of a pavement design is governed by the shear strength of the soil supporting the road, it is usually the geotechnical engineer who tests the soil and determines the pavement design thickness. The transportation engineer often provides design data to the geotechnical engineer, such as the estimated traffic loading, required width of the pavement, and design life of the pavement.

There are many different methods that are used for the design of pavements. Some methods utilize the California Baring Ratio as a measure of the shear strength of the base and subgrade. Numerous charts have also been developed that relate the shear strength of the subgrade and the traffic loads to a recommended pavement thickness (Asphalt Institute 1984). When designing pavements, the geotechnical engineer checks with the local transportation authority for design requirements as well as the local building department or governing agency for possible specifications on the type of method that should be used for the design.

Pavement Types

There are three major types of pavements: (i) flexible or asphalt pavements, (ii) rigid or concrete pavements, and (iii) composite pavements.

Flexible Pavements

Flexible pavements are constructed of bituminous and granular materials. The first asphalt roadway in North America was constructed in about 1870.

Conventional flexible pavements are layered systems with better materials on top where the intensity of stress is high and inferior materials at the bottom where the intensity is low. Adherence to this design principle makes possible the use of local materials and usually results in a most economical design. This is particularly true in regions where high quality materials are expensive but local materials of inferior quality and readily available.

Consider the cross-section of a conventional flexible pavement. Starting from the top, the pavement consists of seal coat, surface asphalt course, tack coat, asphalt binder course, prime coat, granular base course, granular subbase course, compacted subgrade, and native subgrade. In some cases the use of various courses is based on either necessity or economy, and some of the courses may be omitted.

Seal Coat

Seal coat is a thin asphalt surface treatment used to waterproof the surface or to provide skid resistance where the aggregates in the surface course could be polished by traffic and becomes slippery. Depending on the purpose, seal coats may or may not be covered with aggregate.

Surface Coat

The surface course is the top course of an asphalt pavement, sometimes called the wearing course. It is usually constructed of dense graded HMA (hot mix asphalt). It must be tough to resist distortion under traffic and provide a smooth and skid resistant riding surface. It must be waterproof to protect the entire pavement and subgrade from the weakening effect of water. If the above requirements cannot be met, the use of a seal coat is recommended.

Binder Course

The binder course, sometimes called the asphalt base course, is the asphalt layer below the surface course. There are two reasons that a binder course is used in addition to the surface course. First, the HMA is too thick to be compacted in one layer, so it must be placed in two layers. Second the binder course generally consists of larger aggregates and less asphalt and does not require a high quality as the surface course, so replacing a part of the surface course by the binder course results in a more economical design. If the binder course is more than 75 mm (3 in.), it is generally placed in two layers.

Tack Coat and Prime Coat

A tack coat is very light application of asphalt, usually asphalt emulsion diluted with water, used to ensure a bond between the surface being paved and the overlying course. It is important that each layer in an asphalt pavement be bonded to the layer below. Tack coats are also used to bond the asphalt layer to a Portland Cement Concrete (PCC) base or an old asphalt pavement. The three essential requirements of a tack coat are that it must be very thin, it must uniformly cover the entire surface to be paved, and it must be allowed to break or cure before the HMA is laid.

A prime coat is an application of low viscosity cutback asphalt to an absorbent surface, such as an untreated granular base on which an asphalt layer will be placed. Its purpose is to bind the granular base to the asphalt layer. The difference between a tack coat and prime coat is that a tack coat does not require the penetration of asphalt into underlying layer, whereas prime coat penetrates into the underlying layer, plugs the voids, and forms a watertight surface. Although the type and quantity of asphalt used are quite different, both are spray applications.

Base Course and Subbase Course

The base course is the layer of material immediately beneath the surface or binder course. It can be composed of crushed stone, crushed slag, or other untreated or stabilized materials. The subbase course is the layer of material beneath the base course. The reason that two different granular materials are used is for economy. Instead of using the more expensive base course material for the entire layer, local and lower cost materials can be used as a subbase course on top of the subgrade. If the base course is open graded, the subbase course with more fines can serve as a filter between the subgrade and the base course.

Subgrade

The top 150 mm (6 in.) of subgrade should be scarified and compacted to the desirable density near to the optimum moisture content. This compacted subgrade may be the insitu soil or a layer of selected material.

Full Depth Asphalt Pavements

Full-depth asphalt pavements are constructed by placing one or more layers of HMA directly on the subgrade or improved subgrade. In 1960 this concept was conceived by the Asphalt Institute and is generally considered the most cost effective and dependable type of asphalt pavement for heavy traffic. This type of construction is popular in areas where local materials are not available. It is more convenient to purchase only one material, i.e., HMA, rather than several materials from different sources, thus minimizing the administration and equipment costs.

The asphalt base course in the full depth construction is the same as the binder course in conventional pavement. As with conventional pavement, a tack coat must be applied between two asphalt layers to bind them together.

According to the Asphalt Institute 1987, full depth asphalt pavements have the following advantages:

a) They have no permeable granular layers to entrap water and impair performance.

b) Time required for construction is reduced. On widening projects, where adjacent traffic flow must usually be maintained, full depth asphalt can be especially advantageous.

c) When placed in a thick lift of 100 mm (4 in.) or more, construction seasons may be extended.

d) They provide and retain uniformity in the pavement structure.

e) They are less affected by moisture or frost.

f) According to some studies, moisture contents do not build up in subgrades under full depth asphalt pavement structures as they do under pavements with granular bases. Thus, there is little or no reduction in subgrade strength.

Rigid Pavements

Rigid pavements are constructed of Portland Cement Concrete (PCC) and should be analyzed by the plate theory, instead of the layered theory. Plate theory is a simplified version of the layered theory that assumes the concrete slab to be a medium thick plate with a plane before bending which remains a plane after bending. If the wheel load is applied in the interior of a slab, either plate or layered theory can be used and both should yield nearly the same flexural stress or strain. If the wheel load is applied near to the slab edge, say less than 600 mm (2 ft) from the edge, only the plate theory can be used for rigid pavements. The reason that the layered theory is applicable to flexible pavements but not to rigid pavements is that PCC is much stiffer than HMA and distributes the load over a much wider area. Therefore, a distance of 600 mm (2 ft) from the edge is considered to be far in a flexible pavement but not far enough in a rigid pavement. The existence of joints in rigid pavement also makes the layered theory inapplicable.

In contrast to flexible pavements, rigid pavements are placed either directly on the prepared subgrade or on a single layer of granular or stabilized material. Because there is only one layer of material under the concrete and above the subgrade, some call it base course, others a subbase.

Use of PCC Base Course

Early concrete pavements were constructed directly on the subgrade without a base course. As the weight and volume of traffic increased, pumping began to occur, and the use of a granular base course became popular. When pavements are subject to a large number of very heavy wheel loads with free water on top of the base course, even granular material can be eroded by the pulsative action of water. For heavily travelled rigid concrete pavements, the use of a cement treated or asphalt treated base course has become a common practice.

Although the use of base course can reduce the critical stress in the concrete, it is uneconomical to build a base course for the purpose of reducing the concrete stress. Because the strength of concrete is much greater than that of the base course, the same critical stress in the concrete slab can be obtained without a base course by slightly increasing the concrete thickness. The following reasons have been frequently cited for using a base course.

Control of Pumping

Pumping is defined as the ejection of water and subgrade soil through joints and cracks and along the edges of pavements, caused by downward slab movements due to heavy axle loads. The sequence of events leading to pumping includes the creation of void space under the pavement caused by the temperature curling of the slab and the plastic deformation of the subgrade, the entrance of water, the ejection of muddy water, the enlargement of void space, and finally the faulting and cracking of the leading slab ahead of traffic. Pumping occurs under the leading slab when the trailing slab rebounds, which creates a vacuum and sucks the fine material from underneath the leading slab. The corrective measures for the pumping include joint sealing, undersealing with asphalt cements, and mud jacking with soil cement.

Three factors must exist simultaneously to produce pumping:

a) The material under the concrete slab must be saturated with free water. If the material is well drained, no pumping will occur. Therefore, good drainage is one of the most efficient ways to prevent pumping.

b) There must be frequent passage of heavy wheel loads. Pumping will take place only under heavy wheel loads with large slab deflections. Even under heavy loads, pumping will occur only after a large number of load repetitions.

c) The material under the concrete slab must be erodible. The erodibility of a material depends on the hydrodynamic forces created by dynamic action of moving wheel loads. Any untreated granular materials, and even some weakly cemented materials, are erodible because the large hydrodynamic pressure will transport the fine particles in the subbase or subgrade to the surface. These fine particles will go into suspension and cause pumping.

Control of Frost Action

Frost action is detrimental to pavement performance. It results in frost heave, which causes concrete slabs to break and softens the subgrade during the frost melt period. In northern climates, frost heave can reach significant depths. The increase in volume of 9% when water becomes frozen is not the real cause of frost heave.

Frost heave is caused by the formation and continuing expansion of ice lenses. After a period of freezing weather, frost penetrates into the pavement and subgrade.

When water freezes in the larger voids, the amount of liquid water at that point decreases. The moisture deficiency and the lower temperature in the freezing zones increase the capillary tension and induce flow toward the newly formed ice. The adjacent small voids are still unfrozen and act as conduits to deliver the water to the ice. If there is no water table or if the subgrade is above the capillary zone, only scattered and small ice lenses can be formed. If the subgrade is above the frost line and within the capillary fringe of the groundwater table, the capillary tension induced by freezing sucks up water from the water table below. The result is a great increase in the amount of water in the freezing zone and the segregation of water into ice lenses. The amount of heave is at least as much as the combined lens thicknesses.

Three factors must be present simultaneously to produce frost action:

a) The soil within the depth of frost penetration must be frost susceptible. It should be recognized that silt is more frost susceptible than clay because it has both high capillary and high permeability. Although clay has a very high capillarity, its permeability is so low that very little water can be attracted from the water table to form ice lenses during the freezing period. Soils with more than 3% finer than 0.02 mm are generally frost susceptible, except that uniform fine sands with more than 10% finer than 0.02 mm are frost susceptible.

b) There must be a supply of water. A high water table can provide a continuous supply of water to the freezing zone by capillary action. Lowering the water table by subsurface drainage is an effective method to minimize frost action.

c) The temperature must remain freezing for a sufficient period of time. Due to the very low permeability of frost susceptible soils, it takes time for the capillary water to flow from the water table to the location where the ice lenses are formed. A quick freeze does not have sufficient time to form ice lenses of any significant size.

Improvement of Drainage

When the water table is high and close to the ground surface, a base course can raise the pavement to a desirable elevation above the water table. When water seeps through pavement cracks and joints, an open graded base course can carry it away to the road side. Cedergren 1988 recommends the use of an open graded base course under very important pavement to provide an internal drainage system capable of rapidly removing all water that enters. A well coordinated design of the subsurface and surface drainage systems is required.

Control of Shrinkage and Swell

When moisture changes cause the subgrade to shrink and swell, the base course can serve as a surcharge load to reduce the amount of shrinkage and swell. A dense graded or stabilized base course can serve as a waterproofing layer, and an open graded base course can serve as a drainage layer. Thus, the reduction of water entering the subgrade further reduces the shrinkage and swell potential.

Expedition of Construction

A base course can be used as a working platform for heavy construction equipment. Under inclement weather conditions, a base course can keep the surface clean and dry and facilitate the construction work.

Based on the above reasoning, there is always a necessity to build a base course. Consequently, granular or open graded base courses have been widely used for rigid pavements.

Types of Concrete Pavement

Concrete pavements can be classified into four types: jointed plain concrete pavement (JPCP), jointed reinforced concrete pavement (JRCP), continuous reinforced concrete pavement (CRCP), and prestressed concrete pavement (PCP). Except for PCP with lateral prestressing, a longitudinal joint should be installed between two traffic lanes to prevent longitudinal cracking.

Jointed Plain Concrete Pavements

All plain concrete pavements should be constructed with closely spaced contraction joints. Dowels or aggregate interlocks may be used for load transfer across the joints. The practice of using or not using dowels must be considered. The advantages of a joint free design were widely accepted. It was originally reasoned that joints were the weak spots in rigid pavements and that the elimination of joints would decrease the thickness of pavement required. As a result, the thickness of CRCP has been empirically reduced by 25 to 50 mm (1 to 2 in.) or arbitrarily taken as 70 to 80% of the conventional pavement.

The formation of transverse cracks at relatively close intervals is a distinctive characteristic of CRCP. These cracks are held tightly by the reinforcements and should be of no concern as long as they are uniformly spaced. The distress that occurs most frequently in CRCP in punchout at the pavement edge. This type of distress takes place between two parallel random transverse cracks or at the intersection of Y cracks. If failures occur at the pavement edge instead of at the joint, there is no reason for a thinner CRCP to be used. The 1986 ASHATO design guide suggests using the same equation or nomograph for determining the thickness of JRCP and CRCP. However, the recommended load transfer coefficients for CRCP are slightly smaller than those for JPCP or JRCP and so result in a slightly smaller thickness of CRCP. The amount of longitudinal reinforcing steel should be designed to control the spacing and width of cracks and the maximum stress in the steel.

Prestressed Concrete Pavements

Concrete is weak in tension but strong in compression. The thickness of concrete pavement required is governed by its modulus of rupture, which varies with the tensile strength of concrete. The preapplication of a compressive stress to the concrete greatly reduces the tensile stress caused by the traffic loads and thus decreases the thickness of concrete required. The prestressed concrete pavements have less probability of cracking and fewer transverse joints and therefore result in less maintenance and longer pavement life.

Prestressed concrete has been used more frequently for airport pavements than for highway pavements because the saving in thickness for airport pavements is much greater than that for highway pavements. Prestressed concrete pavements are at the experimental stage, and their design arises primarily from the application of experience and engineering judgement.

Composite Pavements

A composite pavement is composed of both HMA and PCC. The use of PCC as a bottom layer and HMA as a top layer results in an ideal pavement with the most desirable characteristics. The PCC provides a strong base and the HMA provides a smooth and non reflective surface. However, this type of pavement is very expensive and is rarely used as a new construction. Composite pavements are mostly the rehabilitation of concrete pavements using asphalt overlays.

Pavement Sections

The design of composite pavements varies a great deal. HMA placed directly on the PCC base, is a more conventional type of construction. A disadvantage of this construction is the occurrence of reflection cracks on the asphalt surface that are due to the joints and cracks in the concrete base. The open graded HMA serves as a buffer to reduce the amount of reflection cracking. Placing thick layers of granular materials between the concrete base and the asphalt layer, can eliminate reflection cracks, but the placement of a stronger concrete base under a weaker granular material can be an ineffective design.

Roads with an inadequate pavement section or weak subgrade are susceptible to bearing capacity failures caused by heavy wheel loads. The heavy wheel loads can cause a general bearing capacity failure or a punching type shear failure. These bearing capacity failures are commonly known as rutting, and they develop when the unpaved road or weak pavement section is unable to support the heavy wheel load.

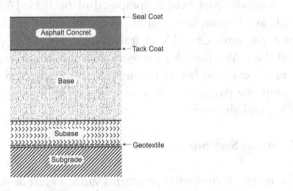

23.1 Typical Section
Asphalt Concrete Pavement

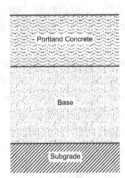

23.2 Typical Section
Portland Concrete Pavement

Chapter 24

GROUND SUBSIDENCE

Among the major geologic and environmental hazards, ground subsidence is as important as landslides, earthquakes, volcanoes and floods. Individual cases of subsidence are usually not as dramatic as a volcanic eruption or a major landslide, and do not often involve loss of life. Subsidence is often a widespread ground hazard. It is not restricted to narrow earthquake belts or steep slopes, but can occur in widely scattered situations, normally where the geological causes behind the potential failures are hidden below the ground surface. In terms of ground engineering subsidence is a subtle, but dangerous threat, and it warrants careful assessment at the geotechnical site investigation stage, especially in certain geological environments.

The types of ground subsidence vary significantly. Slow long term movements, due to sediment dewatering or some methods of mining, can disturb large areas, with gentle but destructive effects. Alternatively, instantaneous ground collapses, such as those where soils drop into buried limestone caves or abandoned mines, can cause total collapse over areas of ground, but usually separated by long time intervals of relative stability. For example old mine openings without backfill have caused sinkholes to develop in urban and rural areas such as Timmins, Canada. Some cases achieve a wide notoriety. The Leaning Tower of Pisa, the Florida sinkhole collapses, the Kidd Creek rock wedge failure, the widespread subsidence in the Central Valley of California, the disappearance of the peatlands and the drowning of Venice are all well known, but each has been caused by a separate composition of geological processes and events.

One way of classifying ground subsidence is to start with a division into endogenic, natural processes and exogenic, man made processes. The purely natural processes, including tectonic movements, volcanic deflation, rock solution and deep sediment compaction, are very slow and of limited impact on most engineering time scales. In contrast, man's activities can induce rapid and serious subsidences, notably through mining, groundwater withdrawal causing clay consolidation, subsurface drilling particularly in wet cohensionless soils, peat wastage, accelerated rock solution and soil collapses. In all these cases, the subsidence is the result of interaction between man's activities and existing geological conditions. It is important to consider the range of geological processes which cause the ground subsidence.

The spectrum of subsidence processes is influenced by man and nature. For building settlement the type and extent are largely the result of the imposed loading, and it has its literature base within the fields of soil mechanics and foundation engineering. For severe cases of settlement the geological conditions are especially relevant to the scale of ground failure. Subsidence in areas of permafrost has important geological background and is often due to man's disturbance and melting of the ground ice beneath and around structures.

The geoengineering specialist is usually at the forefront in dealing with ground subsidence.

Often the major role is played by man in inducing the subsidence of his environment. Natural ground subsidence processes have changed little over the millennia, but the increasing worldwide pressure on the resources of the land mean that ground subsidence is increasing in areas of aggressive development and intense urbanization.

1. Cavern Collapse

Natural voids can occur in a wide variety of geological situations, though their numbers are dominated by solution cavities in limestones. They comprise the aspect of geology less understood by many geoengineers, and site investigations to assess the distribution of concealed voids is difficult and incomplete. Natural cavities are not often found but they can create hazardous ground conditions.

Natural macrocavities rarely occupy more than 1% of a rock mass, though they can be localized into dangerously cavernous horizons or frature zones.

Cavities can be formed by various types of underground mechanical erosions, or alternatively by fissure opening in the head zones of landslides. The most important is underground chemical erosion. Solutional processes can achieve results in the initial stages of development, when weaknesses in the solid rock are limited to microfractures too narrow for mechanical transport. Furthermore, solution is effective at producing large cavities where its effort is concentrated along well defined fracture lines and not diffused through the rock mass, thus the importance of cavities in massive limestones.

Limestone Caves

Caves are formed in strong, massive limestones, with low rock permeability, which contain widely spaced fractures available for solutional enlargement by through flowing groundwater. The same applies to dolomites and marbles. Weaker, more porous limestones have more diffuse groundwater flow and rarely develop large cavities. Caves are rare, in chalk. Tuff and travertine may contain caves at shallow depth where deposition by surface streams encloses voids beneath the projecting shelves of waterfalls.

Caves are identifiers of karst landscapes which are characterized by the underground drainage through the caves. Although developed in a variety of styles, most karst terrains are discovered by their sinkholes, dry valleys, closed basins and rock landscapes. Limestone is such a common rock that a worldwide overview shows few countries without karst and caves. The larger areas of limestone, with the greatest impact on engineering practice, are in southern China and the southeastern USA.

Limestone solution rates are directly related to the amount of rainfall, and also to the levels of groundwater carbon dioxide, most of which is derived from plant and bacterial activity within the soil cover. Consequently the lush vegetation of the wet tropics, in places such as Malaysia, south China and Barbados often, are optimal for cave development. Limestones in these environments are normally more cavernous than similar lithologies in cooler climatic zones. Natural solution rates are so low that approximately 5000 years is the required time to form a limestone cave one metre in diameter. The engineering hazard on limestone is not due to cave growth, but is entirely due to the failure of old caves, or, far more commonly, the failure of soils into old limestone caves to form damaging and life threatening sinkholes.

Foundations on Cavernous Ground

The variations in shapes and sizes of natural cavities are such that generalizations concerning ground treatment are rarely appropriate on a single site, and cavities are best treated on an individual basis. Shallow caves can be cleaned out or collapsed and then backfilled with soil or grout mixtures and much deeper ones below the levels of influence may be left untreated. Smaller cavities at intermediate depths can be tolerated by reinforced strip footings, or spread foundations with reinforced rafts, designed to span any potential collapse, or by subsequent grouting, minipiles, or the like.

Larger cavities at critical depth are generally best treated by being filled with concrete grout. If they are open to the surface, a reinforced concrete cap may be warranted with designs following those for sealing mine shafts. Caves may take enormous quantities of grout, with major potential losses into void extensions. Entering a cave and selectively sealing outlets to void extensions has proved worthwhile before grout filling on some Yugoslavian construction sites. In other cases, forming columns with high viscosity grout can prove economical. Extensive cave development may warrant treatment with perimeter and infill grouting as used in old mine workings.

Piles can fail by punching through an underlying cave roof in misleadingly strong limestone, and micropiles with low individual loadings can offer benefits on cavernous rock. Driven piles generally require at least 3 m of sound rock beneath them, and this may need to be proved by a probe, which can be drilled down between the flanges of an I-section steel pile. Heavily loaded cast piles may require proof of 4 to 5 m of sound rock beneath each one. Cavities below and immediately to the side of a foundation base can offer a hazard, and major buildings on limestone in the Hershey Valley, Pennsylvania, are founded on caissons whose integrity has been checked by drilling vertical holes beneath each, and also inclined holes splayed out at 15°. In northern Greece, a bridge was satisfactorily founded on cast-in-situ piles placed right through a limestone cave; a reinforcing I-beam was placed to vertically span the cave, and concrete was then poured into a canvas casing which expanded to twice the diameter of the pile and set to form a column through the cave. Caves may be spanned with steel piles placed through them and bearing on the floor, although grouting and filling is generally less costly.

Buildings may stand on cavernous ground, with the benefit that some cavernous rocks are reasonably strong so, when isolated cavities are found and then filled or spanned, they provide stable bearing capacity. Cavities in overburden soils have significantly more risk. A single unknown cavity which remains undiscovered, constitutes an engineering hazard totally out of proportion to its size. Thorough site investigation, based on geological awareness, and supplemented by widely based geophysics is therefore of prime importance.

2. Sinkhole on Limestone

Surveys of karst terrains show surface hollows which drain underground, and, other than on a permeable karstic rock or in a desert, would fill up to form lakes. These are known as dolines, closed depressions or sinkholes. In England the word sinkhole generally describes a depression with a visible stream sink in it, however the North American usage of sinkhole covers all such hollows, with or without stream sinks.

Sinkholes will often form on terrains of limestone or dolomite or where either of these rocks occurs not far below the surface. They can form over any rock which is soluble or cavernous. Individual sinkholes may be less than one metre, or more than 100 m, in both depth and diameter, may be circular or elongate, and can have profiles which are conical, cylindrical, saucer shaped or irregular.

Four main types of sinkholes can be recognized.

The solution sinkhole forms by slow surface erosion and is the karst equivalent of a valley. The hazard which it makes to foundation engineering lies in the fact that cavities of some sizes must exist directly or obliquely below it, and these can promote subsidences and other types of sinkholes in its floor.

The collapse sinkhole is created by bedrock failure, and is rare except on a geological time scale.

The buried sinkhole is either of the two previous forms filled and obscured by sediment, and, along with the pinnacled rockhead common on limestone, creates difficult and potentially hazardous ground conditions. The subsidence sinkhole is formed by failure of a soil or weak rock into underlying cavernous limestone. It is the most frequent type, and its rapid development makes it the major engineering hazard.

The distribution of limestone sinkholes is worldwide, with notable concentrations in southeast Asia, the eastern USA, Barbados and parts of

Europe. Their impact on foundation engineering has promoted geoengineering literature, mostly with subsidence sinkholes.

Subsidence Sinkholes

Subsidence or collapse of a soil overburden into the fissures and caves of an underlying limestone creates subsidence sinkholes without involving failure of the rock. The cover material may be alluvium, boulder clay, clay or any other soil, or in some cases may be consolidated rock. The resultant ground hollows, whether formed rapidly or slowly, are known as subsidence sinkholes.

The depth of a subsidence sinkhole is limited to the soil depth, and most stabilize to a near conical shape with the apex close to the limestone fissure below. The width relates to the soil slope stability. It may be over five times the depth in loose sands in Florida, or less than three times the depth in cohesive boulder clay in Ontario or England. Many sinkholes in these soils are less than a metre across, but others over 100 m wide have formed by recent collapse in Alabama and Florida.

Some limestone terrains have a sufficiently high density of sinkholes to influence land use. In parts of Georgia, sinkholes occupy 20% of the ground area. There are estimates of 6500 new sinkholes formed since 1950 in the USA, mostly in the southeast, together with many more unrecorded foundation and structural repairs have cost hundreds of millions of dollars. The frequency of sinkhole collapses in the developing parts of north Florida has had such an impact that insurance companies initiated funding for the Sinkhole Research Institute in Orlando.

The mechanism responsible for subsidence sinkholes is the down washing of sediment by rainwater draining towards a bedrock fissure, though there are variations on this process. The sinkhole develops by periodic collapses, flow or slumping of the soil.

3. Subsidence and Collapse on Chalk

Chalk is a fine grained, soft, pure, white and porous variety of limestone. It has an extensive outcrop across northern Europe, including nearly 15% of the solid geology of England. Chalks occur elsewhere in the world, including the Caribbean, the Middle East and Australia, and though some of these are not of Cretaceous age, they are all geologically young. The distinctive feature of chalk is that, like other limestones, it is slowly dissolved in natural water and can contain cavities, and it has much lower mechanical strength, especially when weathered. Chalk forms its own characteristic style of karst landscape, with rolling hills, dry valleys and underground drainage, and with a noticeable lack of bare rock crags except in undercut sea cliffs.

Over 75% of chalk typically consists of shell structures, known as coccoliths, which are less than 5 μm in diameter. The high porosity of the chalk, ranging from 20 to 50%, is due to the open packing of the coccoliths and also to cavities within them. Chalk's variable physical properties relate to its local history of preconsolidation loading and burial. The weaker beds, notably the Upper Chalk which dominates the outcrop in England, have never had a cover of Tertiary rocks. Joints and fissures in the chalk are irregular, and are widely spaced at depth, but their density increases significantly up through the surface weathered zones.

The unconfined compressive strength of dry unweathered chalk ranges from 5 to 27 MPa. However, the saturated strength is reduced to between 2 and 12 MPa, with most loss exhibited by the very porous Upper Chalk. Due to its fine grain size and high water retention, most chalk is saturated. The high porosity of chalk accounts for its spectacular susceptibility to frost shattering. Natural weathering increases the fracturing and disintegration of the chalk, and creates a surface layer of structureless putty chalk of very low strength. Ground subsidence over chalk is mostly related to either the formation of solution cavities or the failure of putty chalk.

Chalk does not warrant the general description of cavernous because of its high fracture density, high porosity and low strength.

4. Mining Subsidence

Total extraction methods of mining remove all of the mineral present, thereby leaving an unsupported mine roof which is freely allowed to collapse. Consequently, ground subsidence is inevitable and extensive. The costs of surface subsidence damage, to both land and structures, are more than compensated by the increased mineral recovery when no support pillars are left in place, and also by the opportunity to mechanize longwall extraction faces in thin coal seams. The ground subsidence is predictable and rapid, in contrast to the random, lingering hazard of many old pillar-and-stall mines, and it is controllable to the extent that it can be locally reduced in some circumstances.

Underground mining of coal in Britain and part of Europe is dominated by mechanized longwall working, but this is not the only total extraction technique. Much of the mining in North America is achieved by pillar-and-stall working followed by pillar extraction on the retreat. And block caving of various metalliferous ore bodies can cause massive localized surface collapse.

For example a massive rock wedge failure of one the world's deepest copper sulfide mine occurred in Timmins, Canada.

5. Failure of Old Mine Works

Underground mining, both today and in the past, is a major concern in many parts of the world, where its subsurface scale and extent can surprise anyone not directly involved in the mining industry. The removal of mineral ores automatically creates underground voids, which are normally neither feasible nor economic to backfill with waste, and which therefore become problems to surface engineers when subsidence threatens.

The approaches to mining vary significantly. They include total extraction methods, where subsidence is inevitable, but which leads to a new stability. These are dominated by modern coal mining. The alternative method of partial extraction leaves some mineral behind to act as roof support. Mine economics demand maximum extraction and minimal support, which, especially in past eras of short sighted planning, created ground of marginal stability with a long term threat of subsequent failure.

A large proportion of mineral ore bodies are planar sheets – either sedimentary bedded units, or fault-guided vein infills – and their mining style then relates largely to their dip. Vertical or steep veins, especially in strong rock, are worked to leave deep open voids – or stopes – with minimal support of stable walls. Normally a few metres wide, stopes may be hundreds of metres long and deep. Lower angles of dip require progressively more support for the hanging wall or roof, usually in the form of remnant pillars of mineral or perhaps timbers, whose eventual failure will delay ground subsidence for maybe hundreds of years.

Pillar-and-stall mining is the basic method of partial extraction of level or low dip ore bodies. The geometry of the mines varies considerably along with the terminology.

Dimensions in pillar-and-stall mines can vary considerably, and the pillar size is determined by the extraction rate. Extraction of 75% is a common optimum target, creating equal width pillars and stalls in regular stoop-and-room patterns.

6. Salt Subsidence

Salt occurs as a rock and it is highly soluble in water. Consequently groundwater flow in contact with salt causes extensive solution, cavity formation, collapse and ground subsidence. Natural salt subsidence affects large areas over long time spans, but is accelerated and usually localized by artificial brine extraction.

Salt is the mineral halite, NaCl, known as rock salt and it occurs as a sedimentary rock precipitated by solar evaporation of lake or sea water. It is commonly found in the basin environment clay facies of desert redbed sequences, such as the Mercia Mudstones of England. It underlies huge areas in North America such as Goderich, Ontario and is common in sedimentary basins worldwide. It may form relatively pure beds over 100 m thick, but is commonly interbedded with clays to form sequences known as Saliferous Beds.

Salt Solution

A cubic metre of water can dissolve 360 kg of salt, to create a brine with a specific gravity of 1.23. Any assessment of salt solution by groundwater should consider the solution rate and the available initial fracture size as well as the available through flow of fresh water and brine. In real terms a cavity of significant size, nominally a conduit a metre in diameter, can be formed naturally in salt within a few years – in contrast to the time scale of around 10,000 years needed to create a similar cavity in limestone. But large natural cavities are rarely stable in salt, due to the low mechanical strength of both salt (approximate UCS of 10 MPa) and the normally associated clay rocks. Salt solution is usually accompanied by progressive collapse, creating micro-cavernous collapse breccias of insoluable residue, and causing gentle but continuous surface subsidence.

The solution process is rapid such that salt cannot normally exist at the ground surface. The exception is in desert climates, such as at locations within the arid interior zones of Asia. Elsewhere, subsurface salt may only exist by the presence of natural brine springs. Where the geological structure indicates that salt should lie at outcrop, there is usually a zone of collapse breccia, immediately beneath any cover of soil or drift. These breccias may be over 100 m thick, and their evolution may have promoted tens of metres of ground subsidence.

Salt solution may continue beneath the breccia, but the salt itself is largely impermeable. Where surrounded by impermeable clays, ponding of the high density brine on the salt-breccia interface is a natural limit to ongoing solution and subsidence. This restraint is removed where outflow of the brine, and replacement by further fresh water, is possible due either to the regional geology and topography, or, very significantly, to artificial brine pumping. In addition to the regional subsidence over the entire salt outcrop, there is local subsidence where groundwater flow is concentrated in brine streams due to variations in the breccia permeability and where surface soils may slump or collapse into small cavities.

7. Regional Subsidence Due to Groundwater Extraction

Subsidence on clay soils as a direct consequence of the withdrawal or abstraction of groundwater from interbedded sand aquifers is a widespread occurrence. Gentle subsidence bowls develop slowly but can extend over large areas. Their main effects are coastal inundation and deformation of surface drainage gradients, together with casing damage to the wells which initiated the subsidence and some cases of structural damage through ground strain. The Koto area of eastern Tokyo has subsided over 4 m since 1920 due to a pumped water table decline of 60 m. Two million people live below high tide level, and massive flood defences and pumped drainage schemes have been installed, including an extensive landfill project focus to eliminate the flood threat posed by a major typhoon.

This style of induced subsidence has affected areas of Japan, mostly around its coast, together with several areas in the southern U.S.A., and more in geologically young terrains of other countries. On a much smaller scale, the same mechanism can promote destructive settlement around construction sites dewatered by well pointing. This is problematic in normally consolidated clays, where it may require special provision of drainage barriers. Although shallow groundwater abstraction is the main cause of the localized subsidence, an equal amount of ground movement can occur through the abstraction of soil, gas or geothermal fluids, although in these cases the subsequent subsidence may not be everywhere.

The worldwide scale of subsidence damage through groundwater withdrawal is believed to have peaked at this time of unprecedented urban growth and industrialization. There are many case histories of the effects of water table decline in Tokyo which are followed by improved groundwater modelling and ground subsidence monitoring of soil consolidation and soil settlements.

Subsidence Mechanism

Compression of aquifers in direct response to groundwater level lowering has been recognized for several years. It is a simple consequence of the increase in effective stress when the porewater pressure decreases. However, sand is almost incompressible under the stress of shallow aquifers, except for a small amount due to grain rearrangement. Only at the higher pressures such as in deep oil reservoirs does grain fracture cause further compression under increased effective stress. Lithified rock aquifers exhibit only minimal compression on draining, but this can be significant. The Zeuzier Dam, a double arch structure in Switzerland, performed satisfactorily for 20 years, and then, in 1979, developed cracks as it subsided rapidly for 80 mm followed by further slow movement. Although the dam is founded on massive limestone, the subsidence was instigated by considerable under drainage into a road tunnel being constructed 1400 m away and 400 m below the dam crest (Egger, 1983).

The compaction of a sand aquifer in response to groundwater level decline is immediate and elastic, and is usually small. Far more important as a cause of major ground subsidence is the subsequent compaction of interbedded clay aquitards, although not pumped themselves, suffer a similar increase in effective stress as porewater pressures equalize with the lower values induced in the adjacent aquifers. The wide distribution of sand alternating clay layers beneath alluvial plains in parts of the world accounts for this style of subsidence, particularly as the shallow sand aquifers are easy to exploit and over pump.

This consolidation of the clay aquitards is mostly inelastic and non recoverable. The amount of subsidence is a function of the coefficient of volume change, directly related to the hydrologists' nonrecoverable specific storage, which is determined by the existing geology, and also the induced stress change due to water table decline, which is both imposed and controllable by man (Helm, 1984). The specific storage is the amount of water which has to be squeezed out of the aquitard to achieve consolidation at the increased level of effective stress, and this varies with the clay mineralogy. It increases by a factor of three, from the low compressibility kaolinites, through the illites, to the highly compressible montmorillonites. A porosity decrease of a few percent within aquitards is induced by minor drawdown.

For the case of the famous Leaning Tower of Pisa in Italy, the cathedral bell tower is 58 m high, but it is 4 m out of vertical plumb. The tower weighs 14,000 tonnes and it imposes 500 kPa applied stress whereas the allowable bearing pressure is about 50 kPa.

The main settlement is due to consolidation and deformation of the soft clay between depths of 11 m and 22 m. Differential movement probably started due to clay variation in the upper sand and subsequently due to eccentric loading. Fortunately the staged construction allowed progressive soil strength increase, and prevented early failure. Consolidation and settlement is continuing.

The planned stabilization is by controlled induced subsidence of the north side; initially by a 600 tonne weight beside the tower, then by electro osmosis drainage of small boreholes.

8. Consolidation of Clay Soils

The high compressibility of weaker cohesive clay soils mean that they are prone to significant consolidation under an imposed structural load. The direct consequence of loading is consolidation of the clay, as the water content is reduced, and that expulsion of water leads to compaction as the bulk volume decreases. This causes settlement of the structure which imposes the load on the clay. This is a widespread occurrence, prediction and control of which are among the main objectives of our applied science of soil mechanics and foundation engineering. Normal foundation settlement is described in a previous consolidation and settlement section.

Clay consolidation involves a major primary phase of water expulsion and a minor secondary phase of restructuring. The scale of both depends on the geological properties of the clay. Young, organic, montmorillonite rich clays with no preconsolidation history exhibit the greatest compression. Rates of consolidation depend largely on the permeability and thickness of the clays, which determine the rate of drainage.

Structural loading is not the only cause of clay consolidation. On a large scale, slow consolidation is due to the natural overburden load of accumulating sediment, and progressively increases with depth. Sediment profile dating at

Venice shows that mean ground subsidence rates have ranged 0.4-3.0 mm/ year over past millennia.

Diagenesis comprises the chemical and physical processes that turn an unconsolidated sediment into hard sedimentary rock, and it takes place during deep consolidation over geological time. It is responsible for the drained strength increases after primary consolidation is complete. Abnormal groundwater conditions can promote changes detrimental to the soil properties. Through drainage by water undersaturated with silica has transformed kaolinite to gibbsite, with substantial volume loss in some clayey sands in Alabama, causing building settlement and small sinkholes. A similar process is believed to cause large surface collapse on laterite soils in Australia.

Settlement of Structures on Clay

Weaker cohesive clay soils exhibit significantly greater settlements than granular soils, and the movement falls into four components when structural load is applied.

a) Immediate undrained settlement, occurring normally within the construction period, is chiefly due to lateral displacement of the soil, and is notably high in very plastic or organic clays.
b) Undrained creep is the slow continuation of the initial compression; normally it is small and is masked by the increasing rate of consolidation.
c) Consolidation settlement is due to the squeezing out of porewater yielding primary consolidation, and it accounts for most ground movements on clay.
d) Drained creep, also known as secondary consolidation, is caused by the long term restructuring of the clay under uniform stress; and it progressively declines over time.

The factors which determine the amount of settlement are the geology of the clay, both its mineralogy and its diagenetic changes, the preconsolidation history, and the structural load and foundation shape imposed on it. Reference is made to Terzaghi's consolidation theory and his models of one-dimensional compression. Two and three dimensional models, incorporating lateral drainage beneath foundations of small areas, are more difficult to apply, and comparisons of results show that the benefits are limited on a practical scale. Available techniques of settlement prediction often show the final settlement can be predicted to within 20%, but time settlement prediction remains less reliable. One difficulty in prediction arises from the critical importance of the preconsolidation stress, which is not easily determined in the laboratory due to sample disturbance, so full scale field tests and plate loading tests can be beneficial.

In addition to the direct effect of structural loading, clay consolidation can be induced by imposed drainage. On a regional scale, this is mostly due to groundwater abstraction. On a local scale, seasonal variations of water content in clay soils cause shrinkage and heave, creating annual surface oscillations commonly up to 30 mm and locally up to 50 mm in Britain and Canada. Vegetation plays a role with oak and poplar trees especially abstracting large quantities of soil water, and resultant soil shrinkage, settlement and building cracks as noted in sensitive marine clays in Ottawa and Montreal.

The level of acceptable settlement varies considerably with the nature of the structure and its foundations. While 25 to 100 mm is a widely tolerated value, curvature through differential settlement is often more critical, and is normally not acceptable above 1:500. Where direct loading would cause excessive consolidation of a compressible soil, the settlement can be reduced to acceptable levels by raft, piled, spread or compensated foundations. There is available literature on foundation engineering design, and the Latino Americana Tower in Mexico City provides an example of successful foundation design for difficult ground conditions.

9. Subsidence on Peat

Peat is unconsolidated soil consisting largely of plant debris in varying degrees of decomposition. It accumulates in poorly drained areas of upland bog or lowland marshes and swamps where anoxic conditions below the water table create an environment of hydrocarbon preservation. The largest areas of peat bog are in the cold, wet, northern latitudes of Canada, Russia and Alaska, while 30% of Finland is surfaced with it. Peat marshes are well known in the Netherlands and the Fenlands of England, and are extensive in parts of the East Indies and in coastal zones of the southeastern U.S.A. and California.

The high void ratio and extremely low strength of peat create geotechnical engineering difficulties wherever it occurs. Additionally, peat is prone to regional subsidence when drained and worldwide this occurs at rates of 5-100 mm/year, and has locally yielded total subsidence of over 5 m.

Nature and Properties of Peat

Peats are highly variable in origin and nature, and have no universal classification. Peat is defined by its high organic content. Muck can be either the same as peat, or an organic soil with a higher mineral content, or a highly decomposed peat. Both peat and muck classify as histosols in the U.S.A. A mire, or a muskeg, the Canadian term is the system of water, peat and plants whose origins and structure influence the final form of the peat.

Environment divides peats into two major groups. Low moor peat accumulates in the lowland swamps of deltaic or coastal regions often with warm climates. High moor peats are those of upland bogs generally in colder climates. The low moor can be further subdivided (Stephens et al, 1984) into the sedimentary pond peat, the widespread fibrous peats of sedge and reed marshes which form such rich soils, as in the Florida Everglades, and the woody peats left by climax vegetation in swamp forests. The high moor peat of the northern muskegs and upland blanket bogs is formed mainly of mosses, dominated by the extremely spongy sphagnum.

Unfortunately the peat types cannot be related directly to their engineering strengths, as no formal peat classification exists. Of significance is the peat origin and its degree of humification. This ranges on a scale from 1 to 10, from undecomposed plant debris to a totally humified amorphous sludge, with strength and stability decreasing up the scale.

The water content of peat is generally in the range of 500-2000% of dry weight, but it is recorded as high as 3235% and is often as low as 100% in peats above the water table. This water may be in macropores, as in the fibrous peats, or in micropores, as in more humified amorphous peats, and the consolidation profile is significantly influenced by the water distribution. Consolidation theory for peat has been developed. It is complex in its relationship to peat morphology, but it may be verified by field tests. Testing of peat is difficult as laboratory material is always disturbed and unrepresentative, even when collected in large diameter, thin walled soil sampling tubes.

Undrained peat has negligible strength as it acts as a liquid, but drained peat may have an unconfined compressive strength of 20-30 kPa. Consolidated by structural load, the strength rises further, and a safe bearing pressure of 70 kPa has been applied on some Canadian muskeg; however, this value is reduced to 50 kPa on peat over 4.5 m thick, and may be significantly lower on some types of peat. Full depth excavation and backfilling with well compacted select aggregates, or avoiding construction on peat are favoured.

Causes of Subsidence

Subsidence on peat may be induced by loading and/or drainage. The very high water content of peat ensures that consolidation under structural load is large and very rapid in the primary stage. The resultant subsidence is on a scale which can be highly destructive unless appropriate design precautions are used.

Subsidence due to drainage is more widespread. Where highly porous peat, which accumulated and survived in a saturated state, is left above a declining water table, it suffers volume loss due to shrinkage, consolidation, oxidation and erosion. Shrinkage due to desiccation is rapid and nonreversible, normally accounting for a volume loss of 25-45%. Consolidation occurs above and below the water table due to a loss of groundwater support with a declining groundwater level. Elastic compression beneath the water table causes seasonal subsidence and uplift to an extent which depends on both the climate and efficiency of the land drainage with annual oscillations of 50-100 mm and more. Consolidation of surface layers is induced by the tillage which often follows wetland drainage. Biochemical oxidation, due to microbial decomposition, above the lowered water table accounts for most of the long term subsidence of drained peat. It is aided by sporadic burning, either natural or man made, and also wind erosion of dry peat on arable farmland. The permanent loss of peat, by oxidation and erosion, is known as wastage, in contrast to the densification induced by shrinkage and consolidation.

Both the rate of subsidence and the relative roles of oxidation and densification vary with drainage depth, climate and peat morphology. Though the processes overlap in their influence, time generally sees the rapid initial subsidence, due to consolidation, followed by the ongoing wastage.

10. Hydro Compaction of Collapsing Soil

Dry sedimentary materials are prone to internal collapse and subsequent volume loss when water is added to them. These are mainly very fine sand and silt soils, either alluvial or loessic, which have remained dry in semi arid environments. They have porous textures and their strength is derived from intergranular bonds, usually provided by a small percentage of clay when wetted. The failure of these bonds generates soil collapse.

The process is generally known as hydrocompaction, and the materials susceptible to it are referred to as collapsing soils. It is also known as shallow subsidence, but it is always surface related and it can occur to depths of 40 m. The term hydroconsolidation is also used.

Hydrocompaction can promote ground subsidence of up to 5 m over wide areas. The amount of settlement depends on the stress on the soil at the time of wetting. Purely under overburden load, the subsidence normally ranges up to 10% of the thickness of the collapsing soil, and is limited at depth by reduced initial porosity of prewetting below the water table. Some soils, known as conditionally collapsing soils only collapse when wetted under an imposed structural load, and may exhibit volume losses up to 30%.

Subsidence by hydrocompaction is generally caused by manmade activities. These are dominated by land irrigation in arid regions, but also include canal leakage, waste water disposal, other modifications to natural drainage, and even moisture accumulation where soil moisture evaporation is prevented by urbanization. The subsidence can be rapid and destructive, creating ground undulations, fissuring and consequent piping, and severe structural damage, notably to canals, dams, ditches and well casings. On the other hand, gentle water application by irrigation sprinklers causes less damaging subsidence over longer period, and by rehydration it may reduce some forms of soil shrinkage.

Potential hydrocompaction of alluvial or loessic soils may be recognized by saturating samples during laboratory consolidation tests. Upon adding water there is a decrease in void ratio, and consequent subsidence, due to wetting the soils while under an intermediate load increment.

The most widespread collapsible soil is loess, the wind blown silt characteristic of cold continental interiors. Hydrocompaction is exhibited by alluvial silts deposited from mudflows, and is locally recorded on some Aeolian sands, volcanic ashes and colluvial soils. Collapsible soil types can be recognized as a type in any one environment and then predicted on local experience. The arid or semi-arid environment is a prerequisite, though the soils may have been deposited by water and subsequently desiccated. The main areas of hydrocompaction in North America are on the loess of the Missouri basin and on the alluvial soils of California. In Europe, there is collapsible loess in the Danube basin, but the loess of Western Europe has a lower void ratio as it has already collapsed in the modern wetter climate. There are vast areas of collapsible loess in Russia.

Mechanism of Soil Collapse

All collapsing soils are mostly silts or fine sands. Hydrocompaction occurs when soil porosity is greater than 40% and with dry densities ranging from 1.1 to 1.7 t/m^3, which represent porosities of 38-60%. Porosities in excess of 48%, shown by a loose packing with granular contacts, demonstrate the importance of larger voids and bridging clay bonds between the sand or silt grains. Maximum hydrocompaction often occurs in soils with 12% clay, less clay will not support the metastable bond, and more will compensate the collapse by its own expansion.

The hydrocompaction occurs when the addition of water causes dispersion, shearing and failure of the clay bonds. The loss of capillary suction when the soil is saturated promotes the structural collapse, while the low total clay content restricts compensatory expansion. Collapse is rapid where the suction effect is lost in a silt almost devoid of clay, and is normally slower where calcite cements strengthen the clay bonds. Earthquakes are not normally responsible for subsidence on collapsing soils. A block of flats in Ruse, on the Bulgarian side of the Danube, stood on 15 m of loess, and did not move in the 1977 earthquake; but three weeks later it had subsided 250 mm at one end where a fractured water pipe had wetted the loess. The destructive failures of loess in earthquakes, in Central China have been due to landslides.

Above the water table, dry loess may have a safe bearing pressure of 300 kPa or higher. It has sufficient strength to have dwellings cut into its deeper layers in China. However, when wetted hydrocompaction can be initiated under imposed loads ranging from 350 kPa down to nil. Most loess subsidence starts under loads of 70-140 kPa, equivalent to 6-2 m of overburden, thus explaining the dominance of precollapsed soil at depth. The extensive loess of the Missouri basin has undisturbed dry densities of 1.17–1.47 t/m^3, and the low density material compacts up to 20% under load when wetted. The main hydrocompaction hazard is recognized where the dry density is less than 1.28 t/m^3 and the moisture content is less than 10% whereas dense loess over 1.44 t/m^3 or precollapsed material with over 15% moisture can support most structures with only minimal settlement.

Hydrocompaction of Alluvial Silt

An example of hydrocompaction of alluvium is in the San Joaquin Valley of California, where it is superimposed on even greater subsidence profiles due to groundwater abstraction and clay consolidation at depth. In 1915, in the watered ground around a pumping station, hydrocompaction affects over 500 km^2 in an area southwest of Fresno. Subsidence is about 1-2 m, but locally reaches 5 m.

The collapse soils in the Valley are silts in the secondary alluvial fans below the smaller valleys from the Coast Ranges. The primary fans are coarser stream debris where continual fluvial supply never permitted desiccation. Flash floods on the smaller intervening fans created mudflows, which, in the dry climate, were desiccated after deposition and never rewetted before the next flow buried them.

11. Earthquake Subsidence

Deformation of the Earth's crust often involves large scale processes whose monumental size places them beyond the control of man. Fortunately these are extremely slow, and subsidence becomes critical in areas close to sea level. The most dramatic and rapid movements are associated with some earthquakes, and various secondary earthquake effects may involve localized ground subsidence. There are zones of crustal warping which deform by slow, smooth, deep seated, plastic creep and therefore do not create earthquakes. These are referred to as tectonic movements, as they are related to forces within the Earth's interior. Most earthquakes are secondary features of their activity, where sudden stress release creates shock waves and ground vibration.

Tectonic Subsidence

Away from the subduction zones, the main cause of tectonic land subsidence is crustal extension and the consequent thinning. Both the London, UK area and the northern part of Holland are currently subsiding by around 2 mm/year due to crustal deformation of the North Sea basin. This subsidence rate is less than the local rise in sea level – a worldwide occurrence due to the melting of polar ice, but variable in amount due to isostatic deformation of the Earth's crust. Isostasy is the buoyancy concept of blocks of crust floating in equilibrium upon the mantle. Bouyancy shows us why oceanic crust is rarely found on dry land. It is thinner and denser than continental crust, and so it floats lower. The combination of subsiding land and a rising sea level has long term implications for the coastal defences of Holland, and has necessitated improved defences around London, including the Thames Barrier.

Tectonically more active areas, such as Japan, may exhibit higher subsidence rates. There, the coastal region around Nagoya is subsiding at 10 mm/year due to the crustal warping into the offshore basin. Some of the world's major deltas have formed where sediment accretes in a subsiding basin, and continuing crustal sag accounts for a portion of the land subsidence on these deltas. Tectonic subsidence combined with deep sediment compaction beneath the Po delta area at Venice is estimated by researchers to be about 0.4 mm/year and a similar or higher rate pertains beneath the Mississippi delta, although at both sites this is overshadowed by sea level rise, currently about three times that rate, and man induced subsidence.

Isostatic adjustment causes vertical land movements in response to changes in surface load. Perhaps the most notable has been the uplift following the rapid melting of the thick Pleistocene ice caps. The head of the Gulf of Bothnia, in Scandinavia, is rising by over 9 mm/year. Comparable subsidence may be due to loading. Large reservoirs are probably the only man made features heavy enough to induce crustal sag, and Lake Mead, on America's Colorado River, and Lake Kariba, on Africa's Zambezi River, have each created basins subsiding at rates of around 15 mm/year.

Volcanic Deflation

More rapid bedrock subsidence may be associated with movements or magma at shallow depths, particularly beneath active volcanoes. Tumescence or the swelling of volcanic centres, as a result of hot, low density magma rising inside, is common and its monitoring is a technique for the prediction of eruptions. Subsidence occurs during or immediately after the eruption, due to chamber deflation on loss of support by the extruded magma, and at an earlier stage due to sector collapse along grabens. A graben is a block that has dropped down between two normal faults. These events are documented on the Kilauea volcano on Hawaii and on the basaltic and andesitic volcanoes of Japan. Widespread subsidence is rarely more than a metre, and often occurs within a time scale of hours. Its impact is apparent along a coastline, and it is usually overshadowed by the associated volcanic damage.

Earthquake Displacement

The ground movements at the Roman temple in Pozzuoli, Italy have been accompanied by small but locally destructive earth tremors. Similarly, large earthquakes are the result of intermittent fault movement, usually related to deep seated plate activity, and overall surface displacement may have a vertical component involving instantaneous subsidence of large areas. The famous destructive earthquake in Alaska in 1964 was accompanied by over 100,000 km^2 of the coastal region subsiding over a metre, while an adjacent region was uplifted.

12. Hazard Avoidance

The most effective technique of hazard avoidance is an awareness of the geological environment and the implications of it. A base amount of much of the world's land surface has no potential for ground subsidence, but there are specific geological situations where either slow to catastrophic subsidence can take place. Although the statistical chances of ground failure are difficult to quantify in some environments, the accumulated data on these ground conditions can recognize a regional hazard potential.

In areas of development if no appreciation of the subsidence hazard is obtained then the consequence of this can be expensive or a disastrous structural failure. Without awareness, site investigation can become a mindless routine instead of a planned evaluation of any difficult ground conditions which could affect the specific project site. For example, if a site investigation borehole breaches a tunnel due to a survey error, the altered program and findings are useful, for an abandoned railway, but damaging if the track bed is active.

The first stage in subsidence hazard assessment is to identify the significant geological environments. This is normally by recourse to a published geological map, or by reconnaissance geological mapping, as few subsidence hazards have no surface features. It is important to have regional and local knowledge of the soils, rocks and groundwater which could be prone to various types of subsidence. Such information cannot be exhaustive or definitive, and it does not imply that subsidence will occur in these ground conditions. However, it does serve as a reminder that geoengineering works should only proceed on certain types of ground when the possibilities of subsidence have been fully assessed and appropriate precautions have been included in the investigation design and construction stages.

After the preliminary assessment has revealed a potential subsidence hazard, the next stage of investigation depends on the nature of the geology. It may be a field search for natural cavities, a laboratory testing programme on soils due to be drained, or a desk study search for data on past mining. The relevant site assessment approaches are multiple.

A site investigation evolves most economically when it starts with a thorough desk study of available data, and this applies particularly to assessment of both hazards and construction difficulties due to man's past activities. In the developed countries, good geological maps often yield comprehensive general conditions to commence a site investigation, but not enough for detailed design.

Air photographs are widely available, and there are techniques appropriate to identifying sinkholes, peat, gulls, old mines and old shafts, all relevant to subsidence. Records of past mining activity are important to subsidence hazard assessment, but the records are often incomplete and therefore underestimate the subsidence threat. Exhaustive archive searches run up against the law of diminishing returns, and it is generally necessary, as in the case of mining, to assume that all ore seams have been worked until the contrary is proved. Computerized data recording offers benefits in hazard assessment, and will improve further when the backlog of archive data is added to electronic filing systems.

Engineering Geology Maps

Subsidence hazards should be recorded on engineering geology maps, and in some areas constitute a significant proportion of the interpretable data. To cover large areas adequately for planning, maps need to be on scales between 1:10,000 and 1:100,000. There are practical limitations to the production and use of such geologic maps. They are often too specific in the data they encompass. If wider ranging they require more expert geoengineering interpretation. They provide useful planning guidelines, but rarely offer sufficient detail for individual engineering works and final designs for construction or mining purposes.

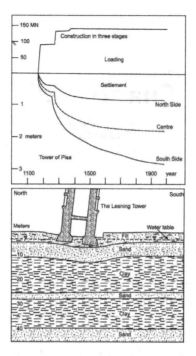

24.1 Leaning Tower of Pisa, Italy

Chapter 25
GROUTING

The use of grouting in geotechnical engineering application has expanded in recent years. Technologies have improved such that grouting can be used successfully in many diverse subsurface applications. Grouting is successfully used in construction and rehabilitation of engineering facilities and treating contaminated soils and rock insitu. The increase in application has led to an increased need for ground engineering technology to verify that the desired ground treatment has been achieved. Testing and investigative tools have improved in grouting technology.

Pressure grouting consists of forcing grout materials under pressure, so as to fill joints and other defects in rock, soil, concrete, masonry, and similar materials. It can modify soil through the filling of voids and pore spaces, or compaction into a denser state. Grouting is used to fill and bond cracks and defects in structural concrete and masonry.

It can be used for cast-in-place concrete where a cementitious grout is pumped into previously placed large aggregate. Specialty geotechnical techniques use pressure grouting as a principal component. These involve a wide spectrum of operations, ranging from the grouting of foundation piles to encapsulation or immobilization of contaminated soil through jet grouting or deep soil mixing with grout. The installation of soil anchors, rock bolts, and foundation piles can be facilitated, and their capacity significantly increased, with supplementary grouting.

The main component of a grout may be cementitious material, a liquid or solid chemical including hot bitumen, or any one of a number of different resins. A combination of two or more components is often used. A wide variety of different filler materials may be included. Grout can have any consistency, ranging from a true fluid to a very stiff mortar like state. It will generally harden at some point after injection, so as to become immobile, and can be designed to have a wide variety of both bond and compressive strengths. It may harden to a very stiff mass or, by design, remain relatively flexible. Grouts will cure into a solid, a flexible gel, or lightweight foam. The application of grout and the use of grouting methods can be advantageous in various earth engineering projects. Grouting is commonly used to achieve any of the following:

- Block the flow of water and reduce seepage
- Strengthen soil, rock, or a combination thereof
- Fill massive voids and sinkholes in soil or rock
- Correct settlement damage to structures
- Form bearing piles
- Support soil and create secant pile walls
- Install and increase the capacity of anchors and tiebacks
- Immobilize hazardous materials and fluids

The beneficial applications of pressure grouting are not limited to the control of water seepage, but also facilitate strengthening and improving virtually all geomaterials.

The early work involved injection of aqueous suspensions or slurries, often containing cement, lime, or clay into joints and seams in bedrock underlying dams in order to reduce water leakage or under seepage leading to problematic piping and soil erosion.

As early practice involved the filling of seams or voids, the grout had to be very flowable, with a maximum particle size considerably smaller than the thickness of the particular discontinuity.

The pore spaces in soils are generally smaller than the apertures of typical rock joints, injection of particulate grout into soils is often of limited success. Accordingly low viscosity chemical solution grouts, which can permeate granular soil formations and then chemically harden, have been developed such as a sodium silicate based formulation, which can be mixed onsite and injected.

The five general types of grouting comprising permeation, compaction, fracture, replacement, and fill are for the main application of grouting in rock, soil and into concrete and masonry structures plus special applications such as grout jacking, grouting in pipes and conduits, leakage control in structures, grouting of underground structures, grouting in extreme environments, explosives in grouting and emergency response grouting.

Grouting Methods

The term grouting is used to describe the process of injecting a material into a geologic formation. The reasons for grouting are significantly influenced by the methods and materials used in a particular application. Soil grouting techniques are usually described as being in one of the following:

1. Compaction grouting
2. Permeation grouting
3. Jet grouting
4. Fracture grouting
5. Soil mixing

Grouting techniques have been used to in support open cut excavation and tunneling activities around the world. They have been widely used in settlement remediation, underpinning, groundwater control, embankment stabilization and providing adequate bearing capacity for structures.

Compaction Grouting

Compaction grout is defined as grout injected with less than 25 mm (1 inch) slump. Normally a soil cement with sufficient silt sizes to provide plasticity together with sufficient sand sizes to develop internal friction. The grout generally does not enter soil pores but remains in a homogeneous mass that gives controlled displacement to compact loose soils, for improved strength gives controlled displacement for lifting of structures, or both. Slump alone is not an adequate measure of injected grout behavior.

Regardless of slump, the formation of grout bulbs as opposed to lenses is enhanced by the use of low mobility grouts.

Compaction grouting involves the use of high pressure positive displacement pumps to inject a very stiff mortar like grout into the subsurface soils.

The purpose of compaction grouting is generally to compact loose soils by radial compression and to carefully lift non settlement sensitive structures and services on the surface, if possible.

Compaction grouting is generally most effective in sandy or silty soils, rubble fills and loose unsaturated sandy clays.

Permeation Grouting

Permeation grouting occurs when grout fills soil pores or rock fissures without causing significant movement or fracturing of the soil or rock formation. It refers to the replacement of water and/or air in the voids between the soil particles or rock surfaces with a grout fluid at pressures designed to prevent fracturing in the grouted soil mass. Permeation grouting has been used as a means for soil stabilization in many engineering projects, where cohesionless soils required an increase in strength or reduction in permeability to be utilized for the intended purpose. It is applicable in coarse to fine sands, depending on the grout mix and soil properties.

Permeation grouting typically requires mixing of the grout components, pumping of the grout to a manifold or injection pipe, and injection through an open ended pipe or sleeve port pipe.

The permeation grout is a fluid of sufficiently low viscosity that it can permeate the porosity of a soil or rock matrix. The fluid grout must be injected at relatively low pressures to prevent hydraulic fracturing of the matrix and undesirable heave of the ground surface.

Permeation grouts can be divided into two categories: chemical grouts (sodium silicate) and cementitious grouts (Portland cement), and often contains additives.

Jet Grouting

Jet grouting is a means of hydraulic cutting and mixing insitu soil materials with a fluid grout injected at high velocity to create a stabilized mixture of soil and grout.

Two specialty components are used to perform this work. A drill manufactured specifically for jet grouting that provides access to insitu soil and consistency during hydraulic cutting and mixing, and a pumping unit to deliver the fluids at the appropriate volumes and pressures.

This method is applicable to a wide range of soil types and has been applied to simple stabilization, underpinning, excavation support, anchors, hazardous waste containment, groundwater control, slope stabilization, erosion protection, and various shaft and tunneling projects.

The typical jetting parameters vary with the three methods. The single fluid system parameters are the water/cement ratio of the grout, grout volume and pressure, number and sizes of nozzles, the drill rod rotation rate and lift speed. Parameters for the double fluid system include air pressure and air flow rate in addition to those for the double system plus water volume and pressure and the number and sizes of water jets. Additionally, in all systems the grout may contain additives, such as bentonite, fluidizers, air and flyash.

The strength of the soil and mixture is vital in underpinning and tunneling applications. The permeability of the soil and grout mixture is the most important on cutoff and containment applications. The primary source of seepage may occur at construction stages, such as the penetration of tieback anchors through cutoff walls.

Soil Fracture Grouting

Fracture grouting is the intentional fracturing of soils using significantly high grout pressures high enough to fracture the soil at the point of injection. The basic concept is to use a grout material that will not permeate the soil. Cement grouts are used frequently and chemical grouts are occasionally used. Fracture grouting is also used in rock grouting.

In fracture grouting, stable fluid grout is injected under high pressure with the intention of introducing controlled fracturing of the ground. Repeated injections after periods of curing tend to densify the adjacent ground, decrease the local permeability, stiffen and strengthen the soil due to the hardened grout lenses, and provide the capability to lift footings or maintain existing elevations during an adjacent excavation.

Because the process requires that the soil be fractured and not necessarily permeated, soil fracture grouting may be used in most stratigraphic types ranging from weak rocks to clays. Due to a broad range of compaction and permeation techniques for coarse grained soils, soil fracture grouting has found special applications in treating cohesive soils.

The largest application for the process has been to remediate against settlement caused by soft ground tunneling under structures. Grout pipes are positioned based on the bearing pressure distribution resulting from the structures to be supported and on local soil stratification.

Ground improvement by soil fracture is based on three mechanisms.

- The soil unit or skeleton is reinforced by a series of hard grout lenses which propagate out from the injection point to form a matrix of hard grout and soil.
- The fluid grout finds and fills voids and causes some compaction in more coarse grained soils along the grout lenses created.
- The plasticity index of saturated clays decreases through the exchange of calcium ions originating from cement or other fillers.

Soil Mixing

Soil mixing is a process by which cementitious materials are mechanically mixed with insitu soil using a hollow stem auger and paddle arrangement. In concept, the process attempts to do by mixing what high pressure jets accomplish during jet grouting. .

The single row multiple shaft auger is generally used for insitu soil mixed walls while double row multiple shaft augers are used for areal treatment of soft or contaminated ground.

As the mixing augers are advanced into the soil, grout is pumped through the hollow stem of the auger shaft and injected into the soil at the tip. The auger flights and mixing paddles blend the soil with the grout.

Insitu soil mixing techniques are applicable to a range of soft alluvial and marine soils, organic soils and reclamation fill. Typically, clay soils with SPT N values of about 4 blows or less and granular soils with N values of about 10 blows or less can be treated.

Soil mixing has been used to address a wide variety of problems, including ground settlement control, slope stability, soil heave prevention, earth retention and liquefaction risk reduction.

Verification Methods

A wide variety of testing methods can be used to either directly or indirectly measure the performance of grouting. These include mechanical, chemical, geophysical, hydraulic and other methods. The focus of these methods is to determine a change in some property of the subsurface after grouting or to detect the presence of grout. Since we cannot see into the grout to know where the grout has gone, or that it has achieved the desired goal, it is possible to use other methods to supplement the construction monitoring. Some of these methods are nonintrusive and/or non destructive and can be used without disturbing or damaging the grouted area. The nondestructive methods generally are indirect and require interpretation of the desired information from some other measured property. Some methods require computer analysis to reduce the data. The more intrusive methods can either collect samples for viewing or other analysis, or measure some insitu properties.

Mechanical Methods: The common approaches are load testing, penetration resistance, static methods, cone penetration test, standard penetration, Shelby Tube, excavation, coring, modulus test, (flat dilatometer and pressuremeter), extensometers (gauge, tape and wire types), tiltmeters, settlement plates, fluid levels, rotating laser levels, density tests, (nuclear method), laboratory shear strength test (unconfined compression, direct shear tests, triaxial compression)

Chemical Methods are also used in verification programs such as:

- pH Indicators
- Total Suspended Solids
- Chemical Dyes

Geophysical Methods:

Geophysical test methods measure physical properties of the ground. Commonly these properties include stress wave propagation velocity, electrical resistance or conductivity, gravitation properties, radar reflectivity, etc. These methods measure properties which can indirectly be related to the action or presence of grout and thus be used for verification of grouting.

Historically, grouters have hidden behind a veil of mysticism, and this continues to a significant degree to this day such as high-tech black magic boxes and claims of proprietary knowledge.

The fact is, grouting is a science, and there is very little special knowledge or ability that is not well established and available. A professional engineer knows that grouting should be approached in a scientific manner and in conformance with engineering practice.

Unfortunately, not all experiences with grouting have proven positive. There have been many instances of poor performance and in some cases damage to the structures or formations that were to be improved. However, this has mostly been attributed to violation of well established basic requirements for proper performance. Some shortcomings are repeatedly experienced, especially injection with little or no advance exploration or engineering knowledge of the site or defects to be corrected. Use of inappropriate grout mixtures, excessively rapid injection rates, and insufficient monitoring and control also contribute greatly. Poor performance is damaging, not only to the affected owner, but also to the reputation of grouting and to all grouting professionals.

It is unfortunate to see grouters who hide behind the veil of false mysticism, claiming to possess proprietary knowledge such that only they are qualified to perform. Worst of all are the large firms that through advertising and other marketing efforts lead unknowledgeable professionals to believe they are the best and represent the state of practice when in reality they are behind and sometimes less competent. When clients receive faulty work from firms they perceive are the best, their perception of grouting suffers.

Verification Program Development

It is important that, prior to the start of grouting, 1) the problem has been defined in detail to the satisfaction of all parties involved, 2) agreement has been reached upon the criteria which define success or failure, and 3) the data which may be necessary for before and after comparisons are gathered prior to the start of grouting.

Prior to the start of a grouting project it is essential to define the existing subsurface problem in sufficient detail so that the condition representing a satisfactory solution to the problem can be specified. It is important that the owner, design engineer, and the grouting contractor agree with the specific purpose of grouting, with the conditions which define success or failure, and with the tests, observations and / or data which will be used to verify the grouting results. It should be understood by all concerned parties that the grout formation, injection rates, pumping pressures and grout takes for each stage of each hole be continuously monitored and that appropriate changes be made during the course of the work, or the work may prove to be a grout disposal project rather than a grouting program.

The subsurface formation into which the grout is injected is opaque to human vision, so the flow of grout into or through the formation cannot be monitored visually. The final location of grout can be controlled to some degree by the practices and procedures used. Viscosity, gel or set time, water cement ratio, jet pressures, rates of injection and other parameters should be set in the design and adjusted during construction to control the rate and distance grout can travel. Since soil and rock are not uniform and variation in them are not always predictable, there always remains some risk that the grout may find its way to some unintended location. The success or failure of a grouting operation cannot be determined until some evidence of performance is obtained.

When the purpose of the grouting is to add strength or stability to the formation, there is generally no way to determine success visually. Often, grouting may be done to increase an existing safety factor which cannot be verified visually. When the grouting operation is completed, records of grout take, setting times, grout pipe location, injection depths and other grouting data are a valuable tool and in some cases maybe strength or indirectly related soil parameters can sometimes be measured at least qualitatively to confirm a change in the soil strength.

Often, it is assumed that if the grout sets in the desired location, the job is successful and the success of the grouting operation is measured by field procedures which verify the grout location. The actual location of solidified grout in a formation can be verified or inferred by various field tests. Specialized equipment and experienced personnel are required.

Unknown or unanticipated underground conditions may adversely affect the grouting performance. A detailed geotechnical site investigation is the best method to avoid unnecessary subsurface surprises. It is unreasonable to assume that any level of investigation will identify every possible situation. When considering a monitoring program, one should consider the possible condition that could affect the grouting. Groundwater may dilute the grout, particularly if grout or ground or groundwater flow rates lead to turbulence.

Fast flowing groundwater could move the grout away from its intended location. Dissolved chemicals in the groundwater may affect.the grout.

Discontinuities in the soil or rock mass could direct the grout to an unplanned location. Soil or rock strata, not encountered in the soil borings, could affect the grout flow and behavior in unexpected ways. Careful monitoring and evaluation of grouting parameters during the course of the fieldwork can allow unknown conditions to be identified and treated appropriately.

When the purpose of grouting is to shut off or divert existing flow or seepage, the success of a grouting operation can be observed visually by observing the change in water flow. When the purpose is to prevent anticipated flow or seepage, success may be inferred at some future date if seepage does not occur. A simple arrangement of wells or piezometers can identify changes in groundwater flows and gradients while the grouting is in progress. For temporary cases, it is necessary for the owner, design engineer, and contractor to agree beforehand upon the amount of seepage reduction or the amount of soil density improvement to support structures, and the targets that constitute success or failure.

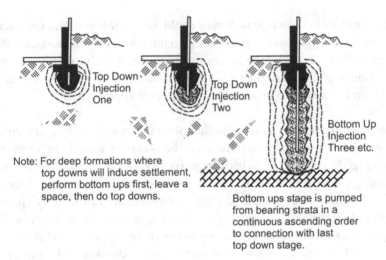

Top Down
Injection
One

Top Down
Injection
Two

Bottom Up
Injection
Three etc.

Note: For deep formations where
top downs will induce settlement,
perform bottom ups first, leave a
space, then do top downs.

Bottom ups stage is pumped
from bearing strata in a
continuous ascending order
to connection with last
top down stage.

25.1 Grouting Procedure, Typical Top Down, Bottom Up

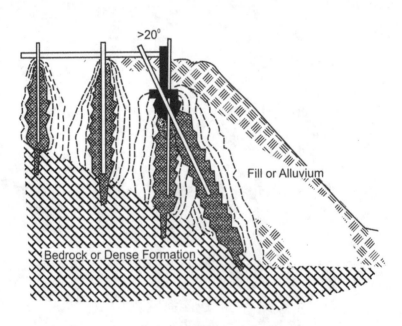

>20°

Fill or Alluvium

Bedrock or Dense Formation

25.2 Angled Drilling and Grouting

Chapter 26

FROZEN GROUND ENGINEERING

Frozen ground in soil or rock exhibits a temperature below 0°C. This definition is based solely on temperature and is independent of the water and ice content of the soil or rock. The large increase in soil strength and the reduction in hydraulic conductivity on freezing has been utilized by geoengineers in the construction of frozen earth structures.

Frost Action In Soil

Frost action in soil is influenced by the initial soil temperature, as well as the air temperature, the intensity and duration of the freeze period, the depth of frost penetration, the depth of the water table, and types of soil and exposure cover. If frost penetrates down to the capillary fringe in fine grained soils, especially silts, then, under certain conditions, lenses of ice may be developed. The formation of such ice lenses may, in turn, cause frost heave and frost boil which may lead to the break up of roads, the lifting of structures or the failure of slopes.

Frost action causes local heaving followed by a loss in the bearing capacity of a soil when thawing takes place. When the thaw occurs the water liberated exceeds that originally present in the soil. A soil usually thaws out from the top downwards, however, it is known for some thawing to take place from the bottom. As the soil thaws downwards the upper layers become saturated, and since water cannot drain through the frozen soil beneath, they may suffer a complete loss of strength.

The following factors are necessary for the occurrence of frost heave, namely, capillary saturation at the beginning of and during the freezing of the soil, a plentiful supply of subsoil water and a soil possessing fairly high capillarity together with moderate permeability. An important factor governing frost heave appears to be grain size since this influences soil moisture movement. For example, well sorted soils in which less than 30% of the particles are of silt size are nonfrost heaving. Frost heave does not occur in clays because of their low permeability. Taber gave an upper size limit of 0.007 mm above which, layers of ice do not develop, although Casagrande suggested that the particle size critical to heave formation is 0.02 mm. If the quantity of such particles in a soil is less than 1%, no heave is expected, but considerable heaving may take place if this amount is over 3% in nonuniform soils and over 10% in very uniform soils. In silts the moisture of the upward capillary rise and/or film flow, if frost penetrates downward into the capillary fringe, forms ice lenses under prolonged and severe freezing conditions. This is because silt particles are small enough to provide a comparatively high capillary rise.

Under the climatic conditions experienced in Britain well drained cohesive soils with a plasticity index exceeding 15% could be looked upon as nonfrost susceptible. Where the drainage is poor and the water table is within 0.6 m of formation level it is suggested that the limiting value of plasticity index should be increased to 20%. The permeability below the frozen zone is principally responsible for controlling heave. The permeability of soft chalk is sufficiently high to permit very serious frost heave but in the harder varieties the lower permeabilities minimize or prevent heaving.

Maximum heaving, does not necessarily occur at the time of maximum depth of the 0°C line, there being a lag between the minimum air temperature prevailing and the maximum penetration of the freeze front. In fact soil freezes at temperatures slightly lower than 0°C. As heaves amounting to 30% of the thickness of the frozen layer have frequently been recorded, moisture, other than that initially present in the frozen layer, must be drawn from below, since water increases in volume by 9% when frozen.

When soil freezes there is an upward transfer of heat from the groundwater towards the area in which freezing is occurring. The thermal energy, initiates an upward migration of moisture within the soil. The moisture in the soil can be moved upwards either in the vapour or liquid phase or by a combination of both. Moisture diffusion by the vapour phase occurs more readily in soils with larger void spaces than in fine grained soils and if a soil is saturated migration in the vapour phase cannot take place.

In a very dense, closely packed soil where the moisture forms uninterrupted films throughout the soil mass, down to the water table, then, depending upon the texture of the soil, the film transport mechanism becomes more effective than the vapour mechanism. The upward movement of moisture due to the film mechanism, in a freezing soil mass, is slow. A considerable amount of moisture can move upwards as a result of this mechanism during the winter. The water table is linked by the films of moisture to the ice lenses.

Before freezing, soil particles develop films of moisture around them due to capillary action. This moisture is drawn from the water table. As the ice lens grows, the suction pressure it develops exceeds that of the capillary attraction of moisture by the soil particles. Moisture moves from the soil to the ice lens. The capillary force continues to draw moisture from the water table and so the process continues.

The amount of segregated ice in a frozen mass of soil depends largely upon the intensity and rate of freezing. When freezing takes place quickly no layers of ice are visible whereas slow freezing produces visible layers of ice of various thicknesses. Ice segregation in soil takes place under cyclic freezing and thawing conditions.

Where there is a likelihood of frost heave occurring it is important to estimate the depth of frost penetration. Once this has been done provision can be made for the installation of adequate insulation or drainage within the soil. Alternatively the amount by which the water table needs to be lowered so that it is not affected by frost penetration can be determined. The base of footings should be placed below the estimated depth of frost penetration, which in the UK is generally at a depth of 0.5 m below ground level, or in southern Canada at a depth of 1.2 to 1.5 m.

Frost susceptible soils may be replaced by free draining gravels. The addition of certain chemicals to soil can reduce its capacity for water absorption and so can influence frost susceptibility. For example, the addition of calcium lignosulphate and sodium tripolyphosphate to silty soils are effective in reducing frost heave. The freezing point of the soil may be lowered by mixing in solutions of calcium chloride or sodium chloride, in concentrations of 0.5 to 3.0% by weight of the soil mixture. The heave of non cohesive soils containing appreciable quantities of fines can be reduced or prevented by the addition of cement or bituminous binders. Addition of cement both reduces the permeability of a soil mass and gives it sufficient tensile strength to prevent the formation of small ice lenses as the freezing isotherm passes through.

Permafrost

Perennially frozen ground or permafrost is characteristic of tundra environments. Permafrost covers 20% of the Earth's land surface, including the Canadian Arctic. During Pleistocene times it was developed over a larger area, evidence of its former presence being found in the form of fossil ground ice wedges, solifluction deposits and stone polygons. The temperature of perennially frozen ground below the depth of seasonal change ranges from slightly less than 0°C to -12°C. Generally the depth of thaw is less the higher the latitude. It is at a minimum in peat or highly organic sediments and increases in clay, silt and sand to a maximum in gravel where it may extend to 2 m depth.

Because permafrost represents an impervious horizon it means that water is prevented from entering the ground so that soils are commonly supersaturated. Hence solifluction deposits and mud flows are typical. Soil material on all slopes greater than 1° to 3° is on the move in summer. Permafrost, by maintaining saturated or supersaturated conditions in surface soils, aids frost stirring, frost splitting and mass wasting processes. Frost action and gravity movements result in certain characteristic features, for example, frost boils, hummocks, and terraces. Annual freezing in permafrost areas brings about changes in surface and groundwater movements which may result in the development of frost blisters, ice mounds, icings or pingos.

There are two methods of construction in permafrost, namely, the passive and the active methods. In the former the frozen ground is not disturbed or is provided with additional insulation so that heat from the structure does not bring about thawing in the ground below, thereby reducing its stability. By contrast the ground is thawed prior to construction in the active method and it is either kept thawed or removed and replaced by materials not affected by frost action.

Permafrost is an important characteristic, although it is not essential to the definition of periglacial conditions, the latter referring to conditions under which frost action is the predominant weathering agent. Superficial structural disturbances have several causes, including frost shattering. Ice wedges may originate as thermal contraction cracks which are enlarged by the growth of ice and subsequently infilled, and in plan may form part of a polygonal network.

Surface cracking may be associated with the formation of stone polygons. Involutions are pockets or tongues of highly disturbed material, generally possessing inferior geotechnical properties, which have been intruded from below into overlying layers due to hydrostatic uplift of water trapped beneath a refreezing surface layer. Frost mounds and pingos are subject to solifluction during their lifetime consequently a depression develops when the ice lens melts. This may be filled with deposits of saturated, normally consolidated fine sediments, which alternate with layers of peat. Mass movements in former periglacial areas may include cambering and valley bulging, landsliding and mudflows.

Geographic Extent and Thermal Regime

Permafrost underlies 20 percent of the world's area, being widespread in North America, Eurasia, and Antarctica. In northern hemispheres it occurs mostly in Canada and the Soviet Union, each country having about one-half of its total land area underlain by it. It is found in most of Alaska, Greenland, northern Scandinavia, and in Outer Mongolia and Manchuria. It occurs at high elevation in mountainous regions in other parts of the world.

Canada has become increasingly aware of her northern frontier. The wealth of natural resources, particularly minerals, will encourage development.

The existence of permafrost usually necessitates modifications in conducting various building development tasks because of the need to counteract its adverse effects. Costs are higher than in temperate regions where there is no permafrost. Increasing knowledge of permafrost properties and characteristics has improved the level of technology to the point where it is possible to build virtually any structure in the permafrost region.

Definition and Origin

The term permafrost is a convenient short form of permanently frozen ground. It is a term used to describe the thermal condition of earth materials, such as soil and rock, when their temperature remains below 0°C (32°F) continuously for a number of years.

Permafrost includes ground that freezes in one winter, and remains frozen through the following summer and into the next winter. This is the minimum limit for the duration of permafrost. It may be only centimetres thick. At the other end of the scale, permafrost is thousands of years old and hundreds of metres thick. The mode of formation of such old and thick permafrost is identical to that of permafrost only one year old and, a few centimetres thick. In this case of the former, even a small negative heat imbalance at the ground surface each year results in a thin layer being added annually to the permafrost. After several thousands of years have elapsed, this process repeated annually can produce a layer of permafrost hundreds of metres thick. However, the permafrost does not increase in thickness indefinitely. Rather a quasi-equilibrium is reached whereby the downward penetration of frozen ground is balanced by heat from the earth's interior. Above the permafrost is a surface layer of soil or rock, called the active layer, which thaws in summer and freezes in winter. Its thickness depends on the same climatic and terrain features that affect the permafrost.

In recent years, it has been realized that permafrost is not necessarily permanent. Changes in climate and terrain can cause the permafrost to thaw and disappear. Thus the term perennially frozen ground is now used instead of permanently frozen ground.

The origin of permafrost is not well understood but it is believed that it first appeared during the cold period at the beginning of the Pleistocene.

For example, it was calculated that at Resolute, NWT, on Cornwallis Island about 10,000 years was required for the ground to freeze to a depth of about 400 m, the present estimated thickness of the permafrost. The temperature of the permafrost in the discontinuous zone at the level of zero annual amplitude generally ranges from a few tenths of a degree below 0°C at the southern limit to 5°C at the boundary of the continuous zone.

The thickest known permafrost in the northern hemisphere is reported to be in eastern Siberia about 900 kilometres northwest of Yakutsk where it has been recorded to a depth of 15000 metres.

In the continuous permafrost zone the mean annual ground temperature decreases steadily from the ground surface to a depth of from 15 to 30 metres. Below this depth the temperature increases steadily under the influence of the heat from the earth's interior.

Whether the permafrost occurs in scattered islands as near the southern limit of the discontinuous zone, it is a factor requiring serious consideration in the development of an area.

It is possible to divide the permafrost region into subregions of varying types of permafrost problems and the methods of coping with them. For example, one well established division is discontinuous and continuous zones. Areas of thawed ground are found in the former zone in contrast to the latter zone where permafrost is found everywhere beneath the ground surface. It is possible to avoid permafrost in the discontinuous zone and employ methods used in temperate regions where it is impossible to avoid permafrost in the continuous zone. Construction and other activities in the discontinuous zone are complicated, however, by the erratic and unpredictable occurrence of permafrost and the proximity of its temperature to 0°C.

In the construction of buildings, for example, permafrost may be avoided in the discontinuous zone and the structure designed for the properties of the soil in the thawed rather than the frozen state. Frozen ground can often be thawed and prevented from reforming by stripping the vegetation and allowing the heat from the building to maintain the soil in the thawed state. Problems can arise, where the permafrost, which is close to 3°C, may thaw slowly during the life span of the structure. In the continuous zone, it is impossible to avoid permafrost and buildings must be designed for the properties of the soil in the frozen state.

Roads and railroads are different because they extend horizontal distances over which the permafrost may change in extent and thickness. In the discontinuous zone, it is possible to avoid permafrost in some areas and design for the properties of the soil in its thawed state. In the continuous zone, the soil is perennially frozen everywhere and construction techniques must be modified accordingly.

Mining presents a different situation because the site of this activity is determined by the location of the orebodies. In the discontinuous zone, an orebody may be either unfrozen, or partly frozen depending on the extent and thickness of permafrost in the vicinity. In the continuous zone, an orebody may lie in or beneath the permafrost depending on the thickness of the frozen ground.

Although there are significant differences between the discontinuous and continuous zones, it is difficult to subdivide these zones. This is because human activities are concentrated at scattered points separated by vast tracts of uninhabited land and the development at any one spot is conditioned by local factors. The method employed at any point of activity is to evaluate the existing natural conditions and use the best possible techniques of site exploration and construction keeping in mind transportation costs, availability of local materials, labour and other cost factors.

It is recognized that kept frozen, soils having high ice contents provide good bearing capacity for structures, but if allowed to thaw, they lose this bearing strength. Structures founded on such soils will be damaged or even destroyed. The location of transportation routes in permafrost regions is governed by the principle to locate the route on well drained soils with low ice contents and to prevent thawing of the base course which causes severe settlement of the grade and makes the route unusable. Mines in permafrost regions face such problems as water seepage into shafts and drifts resulting in massive ice accumulation, and frozen ore which resists blasting and other conventional extraction methods used in temperate areas.

Economic factors and the development of certain practices mean different approaches to the problem of permafrost. For example, in settlements in Canada and Alaska, water and sewage lines are carried in utilidors above ground. In Russia on the other hand, where utilidors are rarely used, one method is to excavate a trench several metres deep, bury the pipe in this trench, backfill with coarse-grained soils which are not frost-susceptible, and heat the water or sewage to a sufficiently high temperature to prevent freezing. Another method is to place the pipes in a precast concrete conduit which is half buried in the ground above the permafrost table. In Canada buildings are mostly one or two storeys high and frequently of wood frame construction on wood pile foundations, in contrast to the Russia where many buildings up to ten storeys are constructed of concrete blocks or precast concrete panels on precast reinforced concrete piles. In Russia large cities with large populations have developed in the permafrost region as a result of widespread exploration of natural resources. The need for large structures to accommodate the various functions of these cities has been recognized even though the cost of construction of such buildings on permafrost is high and greater than in temperate regions. On the other hand, settlements in Canada's permafrost region are small and few in number. The need for large multistorey buildings is not pressing and smaller wood frame buildings are adequate and perform well. Although it is possible to construct large masonry buildings in Canada's North they are not required at this time.

Permafrost is one of many factors which hamper developments in northern Canada. The severe climate and brief summer season drastically shorten the length of the construction and agriculture seasons. Long winters raise annual heating costs. The long period of continual darkness in winter is both a physical and psychological deterrent to any operation. Land transportation routes into the North are limited and water routes are usable for only a few months in the summer. Because of the remoteness of the area from southern regions and the difficulty of marketing exploited resources so as to compete economically with similar ones in temperate regions, resource exploration and exploitation is limited. It is not lack of technical knowledge so much as logistical problems and economic considerations which have caused construction and mining development of the permafrost region of Canada to be slow.

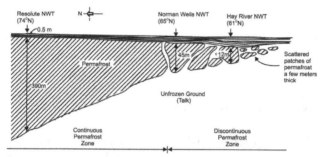

26.1 Permafrost in Cold Regions in Canada,
showing vertical distribution and thickness

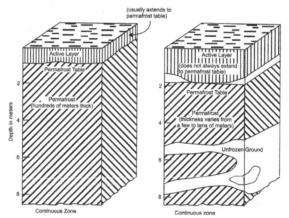

26.2 Permafrost in Cold Regions, showing typical profiles

Chapter 27

FIELD INSTRUMENTATION AND MONITORING

GeoEngineering Instrumentation

The geoengineering practice of instrumentation involves a marriage between the capabilities of measuring instruments and the capabilities of knowledge and experienced technical personnel.

There are two general categories of measuring instruments. The first category is used for insitu determination of soil, rock or groundwater properties, for example, strength, compressibility, permeability and geochemistry normally during the design phase of a project.

The second category is used for monitoring performance, normally during the construction or operation phase of a project, and may involve measurement of groundwater pressure, total stress, deformation, load, strain or contaminant concentrations.

The use of field instrumentation is the selection of instruments with a comprehensive step by step engineering process beginning with a definition of the objective and ending with implementation of the data collection and analysis.

The term geoengineering construction monitoring can be used for construction requiring consideration of the engineering properties of soil, rock and groundwater. The design of geoengineering construction projects is based on judgement in selecting the most probable values within the ranges of possible values for geotechnical properties. As construction progresses and geoengineering conditions are observed or behaviour monitored, the design judgements can be evaluated and, if necessary, updated. Thus, geoengineering observations during construction are often an integral part of the design process, and geoengineering instrumentation is a tool to assist with these important observations.

It is important to plan monitoring programs in a rational and systematic manner. Available methods for monitoring allow us to obtain reliable measurements of these parameters:

- Total stress in soil and stress change in rock
- Groundwater levels and pressures
- Load and strain in structural members
- Deformation
- Temperature
- Geochemical parameters

Reliability is strongly influenced by the extent to which measurements are dependent on local characteristics of the zone in which the instruments are installed. Most measurements of pressure, stress, load, strain, temperature and geochemistry are influenced by conditions within a very small zone and are therefore dependent on local characteristics of that zone. They are often essentially point measurements, subject to any variability in geologic or other characteristics, and may not represent conditions on a larger scale. When this is the case, a large number of measurement points may be required before having confidence in the data.

Geoengineering design and construction will always be subject to uncertainties, and instrumentation will continue to be an important item in our tool box. Several current trends can be identified, each of which will continue in the future and change the state of the practice.

First, there is the advent of automatic data acquisition systems and computerized data processing and presentation procedures.

Second, increasing labor costs in many countries have sharply reduced the availability of competent personnel.

Third, the use of design tools such as finite and boundary element modelling techniques leads to a need for field verification. Geoengineering instrumentation will play an increasing role in providing a check on advanced analytic predictions.

Fourth, there is a trend to develop and improve transducers and to include built-in features that create redundancy and that provide direct output in engineering units.

Fifth, there is a trend toward use of new construction methods. Examples of innovation in the recent past include earth reinforcement, lateral support, and ground modification. These innovations require field verification before they become accepted, and geoengineering instrumentation plays a role.

One of the most important ways geotechnical engineers have to deal with the many uncertainties of site characterization is continued monitoring of subsurface conditions during construction. Often, new information becomes evident during construction particularly if the construction involves making excavations. For example, if a highway cut is to be made into a hillside, geotechnical engineers and engineering geologists base its design on exploratory borings. When the cut is actually made, we examine the newly exposed ground and compare it to the anticipated conditions. If new conditions were found, then the design may need to be changed accordingly.

Another way of dealing with these uncertainties is to conduct full scale tests in the field. For example, these are methods of predicting the load bearing capacity of pile foundations, but often we conduct full scale load tests on piles to verify the computed capacity. Such tests involve installing a production pile at the project site and loading it. These tests should be performed before construction, or at the beginning of construction. If the load test indicates capacities significantly different than those anticipated, then we modify the design, perhaps by adding or deleting piles.

A third way is to install geotechnical instrumentation into the ground. These are devices specifically designed to measure certain attributes in soil or rock. For example, an inclinometer is a geotechnical instrument that measures horizontal movements in the ground. It is prudent to install one or more inclinometers in a slow moving landslide and use the resulting data to help assess the depth and direction of movement, and to judge the effectiveness of stabilization measures.

These techniques of continuing the design process through the construction period are known as the observational method, and form an important part of the geotechnical engineering practice. They represent a significant difference between geotechnical engineering and structural engineering. Infrequently structural engineers need to use these methods because they work with man made materials that are much more predictable and thus do not need such ongoing verification.

Groundwater and Contaminant Monitoring

Groundwater monitoring is a critical component of activities performed at a site where the groundwater may have been adversely impacted by chemicals, i.e. contaminated. The specific purpose of groundwater monitoring will vary depending on the stage of impact identification. The elements of monitoring include:

- initial site characterization using available information;
- refining the site characterization through the installation of boreholes and wells, and sample collection and analysis;
- designing a remedial action plan; and
- implementing and monitoring the remedial action plan.

A well designed sampling and analysis program will:

- provide analytical results of known quality;
- provide a three dimensional characterization of the hydrogeology and chemical presence at the site, including generators, pathways and receptors;
- indicate the extent of temporal variability of the chemical concentrations; and
- stay within resource limitations.

The adequacy of a groundwater monitoring program depends upon:

- an adequate characterization of the site hydrogeology;
- the use of appropriate procedures for the collection and analysis of groundwater samples, and
- statistical interpretation of the results.

Thus, the importance of onsite instrumentation systems and site monitoring observations are essential to a geoengineering project. Inaccurate analytical results cost as much as accurate results initially, but the inaccurate results may return to cause more significant and costly problems. The best approach is to perform an adequate field verification program from the outset to the finish.

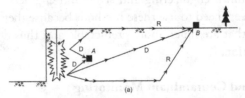

27.1a Idealized Waves for Two Critical Blasting
Geometries, blast transducer geometry

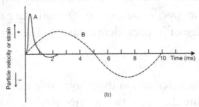

27.1b Idealized Waves for Two Critical Blasting Geometries,
idealized waves D = direct; R = reflected/refracted

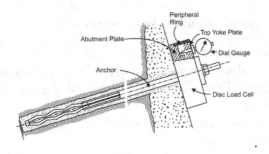

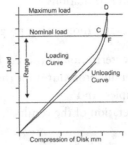

27.2 Disc Load Cell and Calibration Curve for Ground Anchor

Chapter 28
GEOCONSTRUCTION QUALITY ASSURANCE

Engineering construction is one of North America's largest manufacturing industries. Problems in the field may develop from decisions made during the design stage, concerning project drawings and specifications, or during the construction process. Most civil engineering construction projects require multidiscipline teams including onsite quality control and quality assurance personnel.

Ahead of automotive and chemicals, construction represents almost 15% of GNP. Construction is unique in that it is the only manufacturing industry where the factory goes out to the point of sale. Every end product has a life of its own and is different from all others, although component parts may be mass produced or modular.

In civil engineering, mining and earthworks projects success depends on vigilant and continuous onsite engineering inspection and testing to confirm that the materials and workmanship are in accordance with the relevant codes and guidelines, design specifications and construction drawings.

Fulltime quality control and quality assurance has the greatest influence on long term project success. Poor workmanship and poor quality of materials impact the immediate schedule and budget, and they contribute to additional damages, claims and liability in the future, resulting in even higher costs.

During the construction stage a wide variety of laboratory and field testing methods is available for soil, rock, groundwater, aggregates, concrete, asphalt, pavements, steel, roofing and building materials.

Before the contract tendering stages full scale load testing, trial sections, field instrumentation and long term monitoring programs can provide confirmatory findings and design optimization.

Many traditional, advanced and experimental ground improvement methods exist.

Objective

GeoEngineering materials comprising soil, rock, groundwater, aggregates, concrete and asphalt are at the heart of all civil engineering efforts. All the weight of tall buildings or bridges spanning a valley must ultimately be brought to bear on soils and/or bedrock. Failure to realize this simple fact and to provide sound foundation support has caused many problems for engineers, architects, and contractors over the years. Geotechnical materials perform many mundane tasks in engineering such as providing backfill, but they also can be processed to provide important engineering functions such as impermeability for a landfill liner or structural strength for an airfield pavement base course. As the primary constituent of asphaltic concrete and portland cement concrete, they provide us with some of our most economical and high-quality construction materials. It is impossible to visualize any civil engineering effort that is not affected in some manner by geotechnical materials. In engineering design and construction, it is important to be familiar with the properties of geotechnical materials and to be aware of practical techniques for construction with these materials.

GeoEngineering is learned by the practice of the profession and not by education alone. It is an art that develops a concept, conceives a plan or design to bring the concept to fruition, and then builds the concept with full consideration of function, safety, economy, and aesthetics. It is not simply a string of calculations, nor is it a collection of sterile plans and specifications. The miracle of geoengineering is the production of some concrete product for the benefit of people (Rollings 1991). Sustainable development is an important geoengineering consideration.

The art of civil engineering requires applying theoretical tools with practical knowledge of materials, construction, field conditions, and human nature to build a useful and safe structure. Satisfactory results can only be achieved when both theory and practical knowledge are used together. To use one without the other often leads to unsatisfactory and occasionally catastrophic results.

When the natural complexity and uncertainties involved with geotechnical work are factored into a construction problem, the situation often worsens. The late Karl Terzaghi (1961) provided the following commentary on the difficulties of applying purely analytical and rigorous scientific approaches to the art of geotechnical engineering when he observed:

Many problems of structural engineering can be solved solely on the basis of information contained in textbooks, and the designer can start using this information as soon as he has formulated his problem. By contrast, in applied soil mechanics, a large amount of original brain work has to be performed before the procedures described in the textbooks can safely be used. If the engineer in charge of earthwork design does not have the required geological training, imagination and common sense, his knowledge of soil mechanics may do more harm than good. Instead of using soil mechanics he will abuse it.

The need to maintain a balance of sound theory with a practical grasp of design, construction, and material behaviour is essential in civil engineering works and particularly for those using geotechnical materials. Failure to maintain this balance results in costly and expensive failures.

Responsibilities, Risk and Quality Control

At one time, engineers were intimately involved fulltime in all aspects of an engineering project.

Today we have a streamlined approach and have produced many different specialists on projects. The owner, being typically a corporate entity or government agency, has a procurement or engineering staff that will contract with an engineer or architect for design services. The engineer or architect, in turn, often subcontracts work to various specialists and consultants (geotechnical, structural, stormwater control, environmental impact, traffic planning, etc.) and then reassembles this work into a package of plans and specifications with supporting documentation for review by the owner and for any needed permit applications. After revisions mandated by the owner and possibly regulatory agencies, the plans and specifications are ready to send out to contractors for bid.

The owner may use a simple low bid process open to anyone licensed and bonded to do the work, or the bids may be solicited from only the owner's list of approved and prequalified contractors. The first approach is typically required for public agency procurement. The contractor usually has four to six weeks to prepare his bid, and the lowest bidder gets the contract to do the work in accordance with the plans and specifications. This contractor will subcontract out work to other specialty contractors. Inspection of the contractor's work may be done by the owner's staff, the original designer, an independent consultant, or least favourably by the contractor, or it may not be done at all. Intermixed with this process are material and equipment manufacturers who are marketing to owners, engineers, architects, and contractors to have their products specified, approved for use under the specification, or accepted as a substitute for that required in the specification. There are numerous variants to this scenario (construction manager approaches, turnkey construction, design-build, etc.) but our modern engineering construction process is complex with many overlapping areas of responsibility. Unfortunately, the engineer or designer is often removed from what is happening in the field.

After the 1950s, the number of lawsuits involving engineering and construction began increasing and litigation considerations are today a part of every design and construction project. This highly litigious climate has fractured participants into separate and often hostile and suspicious camps centered on the owner, engineer/architect, and contractor. Some steps have been taken to reduce the incidence of litigation and its heavy financial cost, such as contractual agreements to use alternate dispute resolution. However, the current construction climate does not often enough encourage team efforts, and most parties to the construction effort conscientiously try to avoid as much responsibility, risk, and liability as possible.

Probably the largest difference in geotechnical construction work and the remainder of the construction industry is having to deal with the uncertainty of what lies below the surface. Any subsurface drilling program samples only a minuscule fraction of the total underlying materials that may affect the project. Only when excavation or loading reveals the true nature of underlying geomaterials can we be certain of their behaviour, and even then nature may still surprise us. Often the site investigation may be done very superficially, or it is let to the lowest bidder, who may not have sufficient experience or local knowledge to conduct the work adequately. The field information may or may not be made available to the contractor preparing the bid. If it is made available, there is often a specific exculpatory clause releasing the owner from any responsibility for the subsurface conditions and directing the contractor to carry out his or her own subsurface studies. This is often unreasonable because the time the owner allows for bid preparation is usually too short for an adequate site investigation in time to influence the bid preparation, therefore, the contractors should request a time extension to perform all necessary site investigations before construction.

A general legal principal holds that without specific contract provisions to the contrary or unless the owner misrepresents or fails to disclose known subsurface conditions, the contractor bears the risk of unforeseen subsurface conditions. The owner will pay for shifting the risk onto the contractor, and this will be reflected in higher bid prices and increased likelihood of claims during construction. It is more economical in the long term for the owner to shoulder the cost for all comprehensive subsurface investigations for an initial feasibility study and detailed engineering design before construction.

Acceptance of this responsibility typically will require paying for an adequate initial and supplementary site investigations by competent personnel, providing all available information to the bidders, including a reasonable changed conditions clause in the contract, and paying for the designer to monitor the geotechnical work to identify any departures from the design assumptions.

No single document is as important as the specification for construction of a satisfactory project. These specifications should be (1) technically accurate and adequate, (2) definitive and clear, (3) fair and equitable, (4) easy to use, and (5) legally enforceable. They also must be constructable in the real world.

The specifications prepared by the designer engineer or architect will largely determine if interactions and construction on a project will proceed smooth or with difficulty.

There are three obstacles to achieving satisfactory construction once a proper set of specifications is prepared.

a) The low bid owner attempts to gain a final product worth more than he or she is willing to pay.

b) The change order contractor bids low to get the job and then floods the owner with claims for changes.

c) The ivory tower designer treats the specifications as sacrosanct, refuses to visit the site to identify changes from his or her design assumptions, and/or is unable to accept real world construction limitations on the practicality of carrying out a design.

The low bid owner looks constantly for ways to reduce expenditures and may shortchange the initial site investigation or quality of materials used in the work. He or she may fail to call in consultants where needed, not allow or pay for the original designer to monitor construction, and limit or drop requirements for testing. In the end, you will generally get no more than you pay for, and this type of owner will invariably receive poor work.

No perfect set of drawings and specifications has ever been prepared, and the change order contractor can often fabricate alternate interpretations of specification clauses to allow a claim whether legitimate or not. Thankfully, such contractors are not the norm. If encountered, they should be held as rigidly as possible to the specifications (resulting in very unpleasant working conditions) and barred from future jobs where possible.

There are always legitimate changes in a job, however, the contractor should be fairly compensated for these. The ivory tower designer generally has little or narrow experience, which prevents him or her from adjusting designs and specifications for real word conditions or changes in site conditions that depart from the original design assumptions. One also should bear in mind that the job was bid on the specifications, and failure to require compliance with the specifications without valid technical reason unbalances the bidding process and discriminates unfairly against the other bidders who did not get the job but prepared their bid to meet the published specifications. There are times that the requirements of the specification or design concept are critical, and they must be enforced scrupulously, although the importance may not be apparent to the construction and field forces. We see that geoengineering is an art of balance that uses science rather than a rigid theoretical discipline.

The literature on failures in engineered structures commonly calls for increased construction quality assurance inspection to help avoid recurrence of such problems. While there are different philosophies on how inspection may be accomplished. It is crucial that the original designer have significant inspection responsibilities for which he or she is compensated (Rollings 1991). Only the designer is aware of all the nuances and intent in the original design, and it is the designer who is best equipped to identify changes or events that impact these original assumptions and to determine satisfactory solutions to departures from these assumptions. To accomplish this, the designer will need a team of trained and experienced inspectors. Unfortunately, these conditions are often not met in actual construction.

Many aspects of human nature come into play to complicate good interaction between design and construction. The late Karl Terzaghi (1958) described some of this internal conflict as follows:

> *In the realm of earthwork and foundation engineering the absence of continuous and well organized contacts between the design department and the men in charge of the supervision of the construction operations is always objectionable and can even be disastrous…it often depends on whether or not the design and construction departments are on friendly terms with each other. The construction men blame the design personnel for paying no attention to the construction angle of their projects, but they are blissfully unaware of their shortcomings. The design engineers claim that the construction men have no conception of the reasoning behind their design, but they forget that the same end in design can be achieved by various means, some of which can be easily realized in the field, whereas others may be almost impracticable. If conditions are encountered which require local modifications of the original design, the construction engineer may make these changes in accordance with his own judgement, which he believes is sound, although it may be very poor. Important changes of this kind have been made on the job without indicating the change on the field set of construction drawings. The contractor cannot be expected to be interested, or even aware of, the reasoning behind the design. His sole aim is to perform the work covered by the contract at a minimum expense. The inspectors, too, may be inclined to consider uncomfortable items in the specifications as superfluous refinements, conceived in the hothouse atmosphere of the design department. A consultant can never be certain how a structure was built unless he maintains contact with the construction operations. It is the owner's responsibility to ensure that adequate fulltime construction quality assurance inspection and testing are carried out for checking on the quality of materials and workmanship.*

Safety

Construction site safety must be the number one priority when it comes to the general contractor to manage construction operations.

Safety management and safety planning, are required to develop a safe work environment to include, but not be limited, to the following:

- A fully equipped first aid kit should be on site at all times.
- No Trespassing signs should be posted and maintained around the site for the duration of construction.
- Emergency phone numbers for ambulance, fire, hazmat, police, and hospital should be posted at every telephone.
- A list of hazardous materials used on the project should be readily available.
- Hard hats, safety glasses, hearing protection, and other personal safety equipment must be worn by all site staff and visitors.
- Trash and debris must be removed from the construction site on a regular basis.
- All heavy equipment and job vehicles must have backing-up alarm signals.
- All open excavations must be properly protected by barricades and security tape.
- Flagmen, wearing orange safety vests, should be used for traffic control as needed.
- Fire extinguishers and required fire equipment should be certified and maintained.
- Safety meetings should be held on a regular basis with all subcontractors and site staff.
- Power tools should be inspected regularly for defects or unsafe conditions.
- Only qualified personnel should be allowed to operate equipment requiring special training.
- Scaffolding should always be erected correctly and inspected to ensure that it is in good condition.
- Trench shoring should be used to support walls and faces of excavations exceeding five feet.

The most generally applicable legislation is the Occupational Health and Safety Act (OHSA) and the Workplace Hazardous Materials Information System (WHMIS) at various government levels.

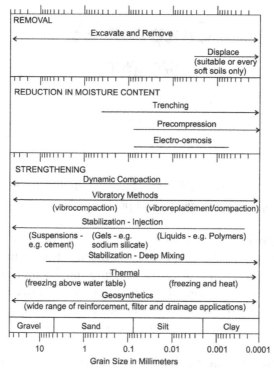

28.1 Conceptual Soil Improvement Methods For Different Soil Ranges Traditional, Advanced and Experimental Approaches

Chapter 29

GEOPHYSICS

Geophysics is the study of the planet Earth excluding the hydrosphere and the atmosphere using methods of physics.

Solid earth geophysics is divided into two main fields of study, global geophysics, and exploration geophysics. The first involves studies of large scale problems relating to the Earth's gross structure and dynamic behaviour. The second deals with applications of geophysical techniques to civil and mining engineering, and environmental investigation for soil, bedrock, groundwater, mineral exploration, petroleum, contaminants, etc.

Geophysical methods are employed by the engineering geologist to obtain data on the subsurface conditions. Use of geophysical techniques involves considerable experience and judgment in the interpretations of results. Common types of geophysical techniques are outlined in this section.

The advances in geophysics have revolutionized our concepts about the Earth. This scientific revolution is largely a consequence of the discovery of seafloor spreading and the subsequent formulation of the theory of plate tectonics. In addition, however, the exploration of the Moon and planets has had a profound impact upon geophysics, as has the development of innovative space technologies, where orbiting satellites have yielded a wealth of global geophysical data. A vital catalyst for these advancements in knowledge has been the growth of computer technology, with the attendant development of powerful new digital techniques for collecting, processing, and analyzing massive quantities of digital data. The overall effect has been to significantly advance our understanding of the Earth and the fundamental way in which we do geophysical research.

Geophysical exploration is important for engineering, geological and environmental (hazardous, toxic and radioactive waste) investigations.

Three classes of objectives are addressed by geophysical surveys (i) the measurement of geologic features, (ii) the insitu determination of engineering properties, and (iii) the detection of hidden cultural features. Geologic features may include faults, bedrock undulations, discontinuities and voids, and groundwater. Earth engineering properties that can be determined insitu include elastic moduli, electrical resistivity and, to a lesser degree, magnetic and density properties. Hidden cultural features available for geophysical detection and characterization include buried underground tanks and pipes, contaminant plumes and landfill boundaries.

Applied geophysics can contribute to the solution of most geotechnical engineering and environmental problems. The geophysical technique does not often directly measure the parameter needed to solve the problem under consideration. Each geophysical procedure measures a contrast. The vast majority of objectives are inferred from the known geologic data and the measured geophysical contrast. Some surveyed contrasts that provide indirect hypotheses are:

- Media velocities from seismic methods to determine overburden thicknesses and the bedrock surface.
- Streaming potentials from the self potential technique to locate a flowing reservoir conduit in a dam abutment.
- High conductivities measured with a terrain conductivity meter to locate an organic plume on the ground surface.
- High apparent conductivity assessed with a metal detector that infers a large metallic cache of possible buried drums.
- Low density contact measured with a gravimeter due to a suspected abandoned shallow mine.

General Observations

Several general observations should be kept in mind when considering applications of geophysical methods.

Resolution, that is the ability of the geophysical measurements to differentiate between two similar geologic solutions, varies widely between geophysical methods. Resolution is a function of time and effort expended and may be improved up to a limit, usually far in excess of the resources available to conduct the study. Ambiguity usually indicates a practical limit on geophysical results before the lack of resolution becomes a factor.

Most geophysical methods do not directly measure the parameter desired by the project geologist, or engineer.

The interpretation of geophysical contrasts is based on geologic assumptions. Ambiguity is inherent in the geophysical interpretation process. The preparation of geophysical models assumes the following:

- Earth materials have distinct subsurface boundaries
- A material is homogeneous i.e. having the same properties throughout
- The unit is isotropic i.e. properties are independent of direction.

These assumptions are often at variance with the reality of geologic occurrences. Ambiguity applies to all geophysical methods, and is most conveniently resolved by understanding geologic reality in the interpretation. Borehole investigations are essential to supplement geophysics. The extent to which these presumptions are valid or the magnitude that the assumptions are in error will have a direct bearing on the conclusions and project plans.

It is important to differentiate between accuracy and precision in geophysical results. Geophysical measurements are very precise. The measurements can be repeated to a remarkable degree on another day, even by another field crew. If accuracy is evaluated as the convergence of the geophysical interpretation with measured geologic data, then geophysical results are not particularly accurate by themselves. However, when appropriate subsurface investigations are integrated with geophysical measurements, large volumes of material can be explored accurately and cost effectively.

There is no substitute for specific geologic or geoengineering observations (such as borings, test pits, trenches, geophysical well logging, and crosshole tests), because of the empirical correlation between results and the inferred objective solution. These borings or other tests are used to validate and calibrate the geophysical results, and ultimately to improve the accuracy of the integrated conclusions.

Interpretation is a continuous process throughout geophysical investigations. The adequacy of the field data to achieve the project objectives is interpreted on site by the field geophysics.

Problems in geological, geotechnical, or geoenvironmental projects require some basic geological information prior to use of geophysical techniques. The determined geophysical contrasts are evaluated and a solution inferred for the likely environment. This approach may require geologic assessment with borings or other field exploration.

Geophysical Methods

Geophysical methods can be classified as active or passive techniques. Active techniques impart some energy of effect into the earth and measure the earth materials' response. Passive measurements record the strengths of various natural fields that are continuous in existence. Active techniques generally produce more accurate results or more detailed solutions due to the ability to control the size and location of the active source.

There are several geophysical techniques that have demonstrated commercial success. There are many surface, subsurface, and airborne geophysical methods used most often or have significant applicability to geoengineering, geoenvironmental, and geologic problems.

Classified by physical effect measured, the surficial techniques are as follows:

- Seismic (sonic) methods.
- Electrical and electromagnetic procedures with natural electrical fields (self potential) resistivity (AC and DC fields), and dielectric constant (radar) theory.
- Gravitational field techniques.
- Magnetic field methods.

Geophysical measures can be applied in the subsurface and above the Earth's surface. Downhole application of geophysics provides insitu measurements adjacent to the borehole or across the medium to the surface. Subsurface applied geophysics gains detailed insight into the adjoining earth materials. Airborne geophysics is usually not as detailed as surface procedures but offers the distinct advantage of rapid coverage without surface contact.

The most commonly used geophysical technique is the seismic refraction method. This method is based on the fact the seismic waves travel at different velocities through different types of earth materials. For example, seismic waves will travel much faster in solid rock than in soft clay. The test method commonly consists of placing a series of geophones in a line on the ground surface. Then a metal plate is placed on the ground surface and in line with the geophones. By striking the metal plate with a sledgehammer, a shock wave (or shot) can be produced. This seismic energy is detected by the geophones and by analyzing the recorded data, the velocity of the seismic wave as it passes through the ground and the depth to bedrock can often be determined.

Seismic Methods

a)   Refraction

The principle is based on time required for seismic waves to travel from source of energy to points on ground surface, as measured by geophones spaced at intervals on a line at the surface. Refraction of seismic waves at the interface between different strata gives a pattern of arrival times at the geophones versus distance to the source of seismic waves. Seismic velocity can be obtained from a single geophone and recorder with the impact of a sledge hammer on a steel plate as a source of seismic waves.

The application is for preliminary site investigation to determine rippability, faulting, and depth to rock and other lower stratum substantially different in wave velocity than the overburden material. Generally it is limited to depths up to 30 m (100 ft) of a single stratum. It is used only where wave velocity in successive layers becomes greater with depth.

b)   High Resolution Reflection

The principle is geophones record travel time for arrival of seismic waves reflected from the interface of adjoining strata.

The application is for determining depths to deep rock strata. Generally applies to depths of over 1000m (3000 ft). Without special signal enhancement techniques, reflected impulses are weak and easily obscured by the direct surface and shallow refraction impulses. This method is useful for locating groundwater.

c)   Vibration

The principle is the travel time of transverse or shear waves generated by a mechanical vibrator consisting of a pair of eccentrically weighted disks is recorded by seismic detectors at specific distances from the vibrator.

The application is velocity of wave travel and natural period of vibration gives some indication of soil type. Travel time plotted as a function of distance indicates depths or thickness of a surface layer. Useful in determining dynamic modulus of subgrade reaction and obtaining information on the natural period of vibration for the design of foundations of vibrating structures.

d)   Uphole, Downhole, and Crosshole Surveys

The principle is:

i)   Uphole or downhole: Geophones on surface, energy source in borehole at various locations starting from borehole bottom. Procedure can be revised with energy source on surface, detectors moved up or down the borehole.
ii)   Downhole: Energy source at the surface (e.g., wooden plank struck by hammer), geophone probe in borehole.
iii)   Crosshole: Energy source in central borehole, detectors in surrounding boreholes.

The application is to obtain dynamic soil properties at very small strains, rock mass quality, and cavity detection. It is unreliable for irregular strata or soft strata with large gravel content. It is also unreliable for velocities decreasing with depth. Crosshole measurements are best suited for insitu modulus determinations.

It is important to note that the seismic refraction method could only be used when the wave velocity is greater in each successively deeper layer. In addition, the seismic refraction method works best when there are large contrasts in materials, for example, soil overlying rock or loose dry sand overlying sand that is saturated by a groundwater table. For inclined strata, only the average depths can be determined and it is necessary to reverse the position of the seismic wave source and geophones and shoot up-dip and down-dip in order to determine the actual depths and the dip of the strata.

The more dense and hard the rock, the higher its seismic wave velocity. The principle can be used to determine whether the underlying rock can be excavated by commonly available equipment, or it is so dense and hard that it must be blasted apart. This information is very important because of the much higher costs and risks associated with blasting rock as compared to using a conventional piece of machinery to rip apart and excavate the rock.

Electrical Methods

a)   Resistivity

The principle is based on the difference in electrical conductivity or resistivity of strata. Resistivity is correlated to material type.

The application is used to determine horizontal extent and depths up to 30 m (100 ft) of subsurface strata. The principal applications are for investigating foundations of dams and other large structures, particularly in exploring granular river channel deposits or bedrock surfaces. It is also used for locating fresh/salt water boundaries.

b)   Drop in Potential

The principal is based on the determination of the drop in electrical potential.

The application is similar to resistivity methods but gives a sharper indication of vertical or steeply inclined boundaries and more accurate depth determinations. It is more susceptible than the resistivity method to surface interference and minor irregularities in surface soils.

c)   E-logs

The principle is based on differences in resistivity and conductivity measured in borings as the probe is lowered or raised.

The application is useful in correlating units between borings, and has been used to correlate materials having similar seismic velocities. Generally it is not suited to civil engineering exploration, but is valuable in geologic investigations.

Magnetic Measurements

The principle is a highly sensitive proton magnetometer is used to measure the Earth's magnetic field at closely spaced stations along a traverse.

The application is difficult to interpret in quantitative terms but indicates the outline of faults, bedrock, buried utilities, or metallic trash in fills.

Gravity Measurements

The principle is based on differences in density of subsurface materials which affects the gravitational field at the various points being investigated.

The application is useful in tracing boundaries of steeply inclined subsurface irregularities such as faults, intrusions, or domes. These methods are not suitable for shallow depth determination, but are useful in regional studies.

There is some application in locating limestone caverns.

The number of geologic issues considered is limited to the problems most commonly encountered in geoengineering or geoenvironmental investigations, since the number of geologic problems is most often larger than the number of geophysical methods. The accompanying matrix the suggested methods versus the problem types, and evaluates the applicability of the method. One cannot rely solely on the applicability of this table, because geology is the most important ingredient of the selection method. This matrix will suggest potential geophysical techniques for particular needs. Geologic input, rock property estimates, modelling, inference effects, and budgetary constraints are the determining factors for the selection of a method. In an attempt to reduce the impact of geology, the evaluation assumes that a moderate degree of geologic knowledge is known before the matrix is used.

A geophysical exploration should be considered early in the development of any site characterization. Monetary and time efficiency will be greatest when the geophysical surveys are part of a phased program, especially at large and/or geologically complex sites. Early geophysical exploration allows some subsequent geologic, engineering, or environmental verification. Problems studied late in the field assessment may have little funding for their resolution remains in budgets to perform the necessary work. Further, there will be little advantage from geophysics performed late in exploration programs, as compared to early geophysical application where subsequent borehole investigations may be optimized in location and detail.

DECISION MATRIX OF SURFICIAL GEOPHYSICAL METHODS FOR SPECIFIC INVESTIGATIONS

	Lithology	Bedrock Surface	Rippability	Detection of Water Surface	Fault Detection	Suspected Voids or Cavity Detection	Insitu Elastic Moduli (velocity)	Material Boundaries, Dip	Linear Subsurface Water Conduits	Landfill Boundaries	Large Ferrous Bodies, Tanks	Conductive Bodies, Ores, Plumes
Seismic Refraction	S	W	W	S	S		W	S				
Seismic Reflection	S	S	S		S	S		W				
SP									W			S
DC Resistivity	S	S		S	S	S		S		W	S	S
Electromagnetic					S			S	W	S	S	W
Ground Penetrating Radar		S		S	S	S		S	S	S	S	
Gravity					S	S		S		S	S	
Magnetics										W	W	W

W - works well in most materials and natural configurations
S - works well under special circumstances of favorable materials or configurations
Blank - not recommended
(U.S. Army Corps of Engineers No. 23, 1998).

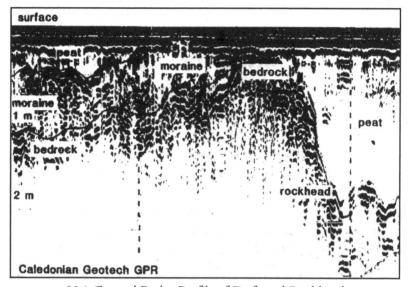

29.1 Ground Radar Profile of Drift and Rockhead

Chapter 30

GEOCHEMISTRY

An increase in human life expectancy and the improved quality of life in recent decades is chiefly due to applied sciences such as chemistry. It is also believed that substances produced synthetically by chemistry are the source of most of our environmental problems.

Geochemistry is the study of the chemical composition of the Earth and its past and current chemical changes. Geochemistry encompasses chemistry and the earth sciences.

Groundwater geochemistry is an interdisciplinary science addressing the chemistry of water in the subsurface environment. The chemical composition of groundwater is the result of the composition of surface water that enters the groundwater reservoir and the reactions with minerals present in the rock that may modify the groundwater composition. Aside from natural processes which affect groundwater quality, recently the effects of pollution, such as nitrate from fertilizers and acid rain, also influence the groundwater chemistry. Due to the long residence time of groundwater in the subsurface environment the effects of pollution may be delayed to become apparent ten to hundreds of years later. An understanding of the groundwater flow and geochemical processes occurring in aquifers is useful to predict the effects of human activities on a broad time scale.

Our interest in groundwater geochemistry is mostly to establish good quality drinking water. Drinking water can be manufactured in a chemical plant, as in coastal areas having desalination plants, however this is costly. The preservation of good groundwater resources has a high priority for most environmental authorities.

The Elements

Four and one half billion years ago, Earth was a ball of molten magma. The magma was made of at least 92 elements, just like our modern day Earth. Elements are made of atoms. Atoms have nuclei of protons and neutrons, with electrons spinning around them. Each element has a different number of electrons and protons in its atoms. An element is a substance that cannot be broken down into anything simpler by natural means, and it is made of only one kind of atom. The simplest and lightest element is hydrogen. It is number one on the periodic chart because it has one proton in its nucleus and one electron orbiting its nucleus. If another proton and electron are added to a hydrogen atom, the number two element on the periodic chart is formed – helium. The more electrons and protons are added to the atoms of an element, the heavier that element becomes. Bismuth, number 83 on the periodic chart, is the heaviest element that is not radioactive with 83 protons in the nucleus and 83 electrons orbiting its nucleus.

Ninety nine percent of the mass of the universe is made of two elements: hydrogen and helium. However, ninety nine percent of the mass of the Earth is made of eight elements: iron, oxygen, silicon, magnesium, sulfur, nickel, calcium and aluminum. The mystery as to why there are so many heavy elements on Earth and only two main light elements in the universe is explained by an understanding of how elements are formed. Some places in the universe have heavy concentrations of hydrogen that are very hot. When hydrogen reaches the temperature of 10 million degrees Kelvin, two atoms of hydrogen can combine to form one atom of helium. When this happens, free neutrons from the nucleus of hydrogen isotopes are released as energy, causing more heat. This in turn causes more hydrogen atoms to fuse together to form helium. When nuclear fusion is taking place this quickly in a cloud of hydrogen in space, so much energy is released that a star is born. Sometimes the nuclear reactions inside a star become so intense and uncontrollable that the star explodes like a gigantic nuclear bomb. This is known as a supernova. It is estimated that a supernova occurs about every 30 years in our own Milky Way Galaxy.

The composition of materials on the Earth's crust is variable. It is important to determine the average composition of the Earth and of the solar system because the relative abundances of elements establish boundary conditions for the creation of the elements and the evolution of the solar system. In chemistry the periodic table of elements is an arrangement of the chemical elements by atomic number with an increasing number of protons, to show the relationship between elements. As mentioned each element is a chemical substance that cannot be broken down to simpler units without changing its chemical properties. The atoms of an element all have equal numbers of protons and electrons. The periodic table of known elements also includes oxygen, carbon, nitrogen, iron, silver and gold.

The average composition of the Earth as a whole is believed to be similar to the composition of the Sun and to the composition of some meteorites. Scientists learn about the formation of the Earth by studying the elemental composition of meteorites which occasionally break through our atmosphere and hit the ground surface.

Lithosphere

The study of the chemical composition of the silicate rocks of the dry land portions of the Earth's surface are important in geochemistry. Petrology and mineralogy examine minerals found in rocks and the relationship to the possible geologic conditions of the formation of rocks and the Earth's crust. Geochemists investigate the distribution of natural elements with similar chemical properties in rocks and minerals and the bearing on geologic processes.

Exchangeable Reservoirs

The atmosphere, oceans, and upper layers of marine sediments, together with the inorganic and biological material of the land areas of the earth, comprise a system where the transfer of material from one area to another takes place rapidly enough to be observed by geochemistry. Interesting Earth characteristics are found where detailed observations of nature and geologic changes are explained by their chemical properties.

The Oceans. The chemical analysis of seawater for all elements in the periodic table is incomplete. In spite of the lack of information on element concentrations in seawater, it is known that the residence time of most elements in the oceans, between supply by weathering of rocks and removal by sedimentation, is short compared to geologic time. Sodium appears to not be removed easily by sedimentation processes, and its concentration is likely increasing gradually with the passage of geologic time. Chlorine is overly abundant in the sea such that it could not be supplied at the present weathering rate during the history of the Earth. The addition of chlorine to the sea must have occurred earlier possibly by volcanic activity in the primitive Earth.

The Atmosphere. The atmosphere is believed to be uniform in composition throughout the troposphere, stratosphere, and mesosphere, up to an altitude of 80 km, with approximately 78.1% N_2, 20.9% O_2, 0.9% Ar, and less than 1% CO_2. The concentration of water vapour may be as high as 3% in saturated air at room temperature or as low as 0.01% or less in very cold or very dry areas. Vaporization and condensation of water vapor underlies weather changes and is generated by adiabatic temperature increase and decrease due to falling and rising air masses. Meteorology is the detailed study of weather, placing special emphasis on thermodynamics. There is attention with microscopic solid particles in the atmosphere, including the migration of radioactive debris from nuclear weapons tests, the behaviour of industrial pollutants, and the relation between natural aerosols and precipitation.

The Sediments. Sedimentary geochemistry is inorganic and it includes investigation into organic chemistry and biochemistry occurring under natural conditions. It includes the formation of valuable mineral deposits, petroleum and coal. Molecular biology with geochemistry has investigated the natural conditions leading to the origin and evolution of life on Earth. Sedimentary geochemistry applies chemical principles and techniques to decode the history of sedimentary rocks including the formation of valuable mineral deposits in mining.

Geochemical Investigation Methods

In principle, all the methods of chemistry and of geology, oceanography, and meteorology are valuable in a geochemical site investigation.

Geochemical findings come from analytical laboratory testing which is an important tool in geological and environmental engineering. It is used in studies of water, soil, and air quality, of formation of rocks and minerals, of fossilization mechanisms, of metal accumulation in organisms from contaminated water and soil, and in determining the suitability of earth materials for civil engineering and mining purposes.

Gas source and solid source mass spectrometry is used for the age determination of rocks and minerals, the temperature history of rock and meteorites, and temperature changes in the seas during ice ages and climate changes.

Several analytical methods for determining element concentrations with accuracy and sensitivity are used in geochemical analysis. They include Atomic Absorption Spectrometry (AAS), X-Ray Fluorescene Spectrometry (XRF), Inductively Coupled Plasma Atomic Emission Spectrometry (ICP-AES), ICP-Mass Spectrometry (ICP-MS), Instrumental Neutron Activation Analysis (INAA) Thermal Ionization Mass Spectrometry (TIMS), and Secondary Ion Mass Spectrometry (SIMS).

Rapid silicate analysis for the major elements includes colorimetric and flame photometric procedures. Chromatography is a tool whereby organic compounds are simultaneously ionized and fragmented in a vacuum by means of electron bombardment. Each compound has a unique spectrum, providing identification. Emission spectrography, x-ray fluorescence, and electron probe microanalysis have contributed to petrology.

Ionization. The manner in which the elements have distributed themselves in our planet depends mainly on their respective abundances and on their chemical properties. Their chemical properties depend in turn on certain fundamental atomic properties; notably, state of oxidation (valence), size (radius), the magnitude and nature of the forces associated with the atom, and finally, a variety of other properties such as polarizability, shape (orbital electron distribution), and whether or not the atom has splitting properties.

The role of a geochemist is to investigate nature with a hypothesis for the cause of what is being observed and relate the observations quantitatively to known chemical properties.

Surface Geochemistry determines the distribution and movements of elements in different parts of the earth (crust, mantle, hydrosphere, etc.) and in minerals to determine the underlying systems of distribution and movement.

Isotope Geochemistry studies the relative and absolute concentrations of the elements and their isotopes in the Earth. Variations in these isotopes reveal information about the age of a rock or the source of the air or water. Isotope geochemistry is divided into two branches – stable and radiogenic.

Cosmochemistry studies the chemistry of the universe, primarily inferences on pre-solar system events, solar nebular processes, and early planetary processes as deduced from minerals in meteorites and from chemical and isotopic compositions of meteorites and their parts. It overlaps geochemistry, with geology, oceanography and meteorology.

Biogeochemistry is the study of the chemical, physical, geological and/or biological processes and reactions that affect the composition of the natural environment.

Organic Geochemistry studies the abundance and composition of naturally occurring organic substances and their impacts on the Earth.

Water Geochemistry is the study of chemical interactions between water and soil/aquifer material that determine the composition of natural water.

Exploration Geochemistry determines the presence of unrefined petroleum, both surface and subsurface. Light gases (methane, ethane, propane, and butanes) present in near surface seeps could be related directly to the type of hydrocarbons reservoired within the producing field. This measurement could be used to predict the oil versus gas potential in advance of subsurface drilling.

Forensic Geochemistry is the scientific-technical discipline in environmental sciences that evolved from exploration and organic geochemistry. Exploration geochemistry reconstructs the history and genesis of natural deposits at depth using geochemical methods and procedures. Forensic geochemistry studies fossil fuel exploration in the attempt to locate and assess man made contamination plumes in soils and groundwater.

Groundwater Quality

The major use of groundwater analyses is to produce information concerning the water quality. Maps displaying the regional distribution of water quality using surveys are carried out routinely for regional well waters.

Groundwater geochemistry may trace the origins and history of water. Water compositions change through reactions with the environment, and water quality provides information about the environments through which the water has flowed. This includes residence times, flow paths and aquifer characteristics, a need which arises from man made pollution sources. Water chemistry can provide useful information since chemical reactions are time and space dependent.

The common parameters used in groundwater chemistry are as follows:

Hardness: ions which can precipitate as hard particles from water
Colour: measured by comparison with a solution of cobalt and
               platinum
Eh: redox potential
Alkalinity: acid neutralizing capacity
Acidity: base neutralizing capacity
TIC: total inorganic carbon
TOC: total organic carbon
COD: chemical oxygen demand
BOD: biological oxygen demand

Standards for Drinking Water

The composition of drinking water has an important health aspect. The standards of the World Health Organization (WHO), and most national authorities, use maximum admissible concentrations. The limits are for constituents which commonly occur, either naturally or due to pollution or water treatment and water supply systems. Too high concentrations of specified elements can limit the use of groundwater for drinking purposes. For example, too high fluoride intake leads to painful skeleton deformations termed fluorosis. It is a common disease in East African Rift Valley countries where volcanic sources of fluouride are present, and in India and West Africa. Elevated arsenic concentrations are associated with sediments partially derived from volcanic rocks of intermediate to acidic composition. Arsenic is associated with sulfides, and high As concentrations can be found in water that contains appreciable (Fe) iron concentrations. It gives rise to black foot disease, visible in a blackening of finger and toe tops, and induces a general lethargy in humans.

Standards for the composition of drinking water and the contribution of drinking water to the intake of elements in nutrition are approximately as follows:

Constituent	Contribution to mineral nutrition (%)	Highest admissible concentrations (mg/l)	Comment
Mg^{2+}	3-10	50	Mg/SO_4 diarrhea
Na^+	1-4	175	
Cl^-	2-15	300	taste; safe < 600 mg/l
SO_4	250	diarrhea	
NO_3	50	blue baby disease	
NO_2	0.1		
F^-	10-50	1.5	lower at high water consumption
As	ca.30	0.05	black foot disease
Al	..	0.2	acidification/Al-flocculation
Cu	6-10	0.1	3 mg/l in new piping systems
Zn	negligible	0.1	5 mg/l in new piping systems
Cd	..	0.005	
Pb	..	0.05	
Cr	20-30	0.05	

Sampling of Groundwater

Groundwater sampling and analysis is carried out by drilling boreholes with sealed monitoring wells. It is important to evaluate which data are required to solve a specific water quality problem. For the interpretation of geochemical analyses, the procedures used to collect and analyze the groundwater samples should be considered.

Procedures for Sampling of Groundwater

A common concern is contamination and disturbance of natural conditions caused by borehole drilling operations. Materials are introduced to the aquifer, comprising drilling fluids, gravel or pack casing materials. It may take considerable time before the influence of such disturbances on groundwater quality is adequately diminished to allow representative sampling in the borehole monitoring well. The time needed to obtain representative samples depends on the groundwater flow rate, the ion exchange capacity, and it may be two to three months when the pumping activity on the screen is limited.

It is known that boreholes and wells which have been out of production for considerable time may yield a water chemistry which is different during production. The main reason is the presence of stagnant water above the well screen, consequently, it is necessary to empty the well for a number of volumes. Excessive pumping or over pumping may draw waters with different composition towards the well screen and cause mixing of waters. A balance has to be found between 2 and 10 well volumes to be purged, and depending on local hydrological conditions.

To accomplish effective flushing, the well should be pumped from just below the air water interface. A favourable method to evaluate the degree of flushing needed is to monitor field parameters such as the electrical conductivity (EC), or pH over time. Stationary conditions may be obtained after emptying a few well volumes. Problems with sampling from existing monitoring wells can be related to faulty well completion, and these are not easy to detect.

Chemical Analysis of Groundwater

A perfect chemical water analysis of groundwater carried out by routine procedures is not likely possible. Both sampling and analytical procedures should be reviewed to obtain the best possible results for the selected parameters in the field studies.

In most cases water analyses are carried out by standard procedures such as described in field hydrogeology handbooks.

When a groundwater sample is brought to the surface, it is exposed to physico-chemical conditions which are different from those in the aquifer. Therefore, measures are needed to prevent changes in the chemical composition of the sample before analysis. Such measures are sample conservation and field verification measurements.

GeoEnvironmental Applications

In environmental geochemistry the chemical principles are applied for predicting the distribution for organic and inorganic pollutants at the earth's surface and in the atmosphere.

Environmental geochemistry establishes and explains the links between the chemical composition of soils, rocks and groundwater and the health of plants, animals and humans. Bedrock geochemistry impacts the composition of soil and groundwater. Pollution, arising from the extraction and use of mineral resources, and industrial spills, distorts natural geochemical systems. Geochemical surveys of soil, rock, groundwater and plants show how major and trace elements are distributed in the Earth's crust, particularly with localized anomalies and contaminant spills.

Subsequent to geoenvironmental site investigations, a subsurface characterization using geochemical testing of samples, followed by toxicological studies, can show the possibility of causal links between the contaminant generator, pathways and human receptors, if any. These have application in environmental risk, environmental impact and environmental site assessments plus followup site remediation measures.

Human health can be affected by a variety of contaminant substances. The exposure of children to low levels of the heavy metal lead can lead to small but measurable decreases in their IQ. The mortality rate of adults appears to be increased by their inhalation of pollutant particles of submicroscopic size suspended in the air of some cities. It is characteristic of chemical related environmental problems that emphasis has shifted from instances of gross pollution with obvious consequences to human health and to the environment, to ones involving widespread contamination at low levels, the effects of which on the large numbers of people affected is a controversial matter.

Earlier it was believed that chemicals emitted into the environment would be assimilated by nature, that is, either the system would convert them into harmless, naturally occurring substances, or the chemicals would be diluted to such an extent they would pose no threat to life. A strategy that the solution to pollution is dilution can be successful for many pollutants. However, it has become clear that many synthetic chemicals are not assimilated because they are persistent. They are unaltered by the action of light, water, air or microorganisms, which often serve to break down, or degrade, many pollutants for very long periods of time. Examples of these persistent substances include pesticides such as DDT, the refrigerants called CFCs, the gas carbon dioxide, and toxic forms of the element mercury.

Acid Mine Drainage

The waters draining from most coal mines are strongly acidic due to oxidation of pyrite in the coal. Waters associated with the mining of sulfide ores are commonly acidic as a result of oxidation of sulfide minerals. Acid mine drainage occurs in mountainous mining areas and from mine tailings disposal. When streams become contaminated by acid mine drainage, adjacent vegetation dies and precipitation of ferric hydroxide occurs over considerable distances. In the absence of mining, acid waters are uncommon because dissolved oxygen in the groundwater is insufficient to produce acidity of the groundwater.

The main strategies to control acid generation are prevention or minimization of water circulation through acid generating material by covering it with an impermeable cap where clayey materials are obtainable. A native soil may be sufficient in arid climates. An alternative is to dispose of acid generating materials under water, as in a flooded mine pit. The treatment of acid mine drainage by placing limestone boulders in affected water courses is not often successful as the boulders become armoured by iron and aluminum that precipitates as the water is neutralized.

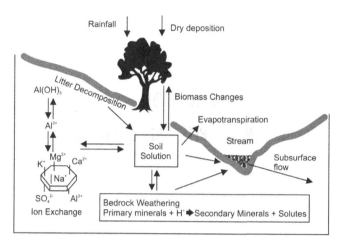

30.1 Geochemistry of Natural Groundwater

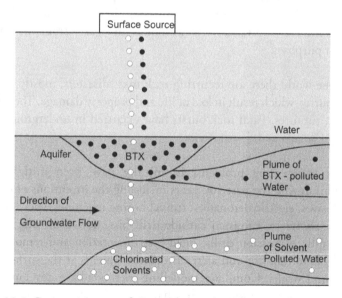

30.2 Contamination of Groundwater by Organic Chemicals
LNAPL = Light Non Aqueous Phase Liquid (floats)
DNAPL = Dense Non Aqueous Phase Liquid (sinks)

Chapter 31
MINING, PITS AND QUARRIES

MINING

Mining and agriculture rank together as the primary or basic industries of human civilization. Mining encompasses the extraction of all natural occurring mineral substances – solid, liquid and gas – from the Earth for utilitarian purposes.

Around the world there are recurring rock mass disasters, mostly landslides and rockbursts, which result in loss of life and property damage. These include large dam ruptures. Fatal rock bursts have occurred in underground mines including Canada and South Africa.

The rock engineering problems in mining have been mitigated with innovations during the past 100 years including the inventions of dynamite and explosives, electric detonators, tunnel boring machines, sprayed mortar, rockbolts, shotcrete, tungsten carbide drill bits, new tunneling methods, and hydraulic percussion drills. Field instrumentation and remote sensing are available to monitor and alert of rock movements at the surface and in underground works. Computer modelling methods have developed to the stage where rock conditions can be characterized and predicted with more accuracy.

The basic industry definitions are:

Mine: an excavation made in the earth to extract minerals
Mining: the activity, occupation, and industry concerned with the extraction of minerals
Mining engineering: the art and science applied to the processes of mining and the operation of mines

Some terms distinguish various materials mined. Geologically, one distinguishes between the following:

Mineral: a naturally occurring substance, usually inorganic, having a definite chemical composition and distinctive physical characteristics
Rock: an assemblage of minerals

Economically, the distinction made between minerals is the following:

Ore: mineral that has sufficient utility and value to be extracted at a profit
Waste or gangue: mineral that lacks utility and value when mined (gangue is more associated with ore than is waste)

When both geologic and economic relationships are involved, the distinguishing terms are the following:

Mineral deposit: geologic occurrence of minerals in relatively concentrated form
Ore deposit: economic occurrence of minerals that can be extracted at a profit

A division of commercial minerals into three main categories, is made on the basis of primary constituent and usage. Metallic ores include ores of the ferrous metals (iron, manganese, molybdenum, and tungsten); base metals (copper, lead, zinc, and tin); precious metals (gold, silver, and platinum); and radioactive metals (uranium, thorium, and radium). Non metallic ores consist of industrial minerals such as phosphate, potash, stone, sand, gravel, sulfur, salt, and industrial diamonds. The mineral fuels or fossil fuels, include coal, petroleum, natural gas, uranium and less common sources (lignite, oil shale, tar sand, and coal bed methane). The activity associated with the extraction of petroleum and natural gas has developed into a separate industry with a specialized technology.

The essence of mining in extracting mineral from the Earth is to drive or construct an excavation, as a means of entry, from the ground surface to the mineral deposit. When the value of the mineral is established with confidence by drilling and sampling, usage of the terms ore and ore deposit is preferred. If the excavation is entirely open to or operated from the surface, it is called a surface mine or open pit mine. If the excavation consists of openings for human entry driven below the surface, then it is an underground mine. The specific details of procedures, layout, and equipment used distinguish the mining method, uniquely determined by the physical, geologic, environmental, economic, and legal circumstances that prevail.

Open pit and underground mines plus tailings disposal methods require comprehensive geotechnical, geomechanics and geoenvironmental input through the feasibility, planning, design, construction, operation and closure stages.

The application of rock mechanics principles in underground mine engineering is based on simple premises. Firstly, it is postulated that a rock mass can be characterized by a set of geomechanical properties which can be measured in standard tests. Secondly, it is asserted that the process of underground mining generates a rock structure consisting of voids, support elements and abutments, and that the geomechanical performance of the structure can be analyzed using the principles of engineering mechanics. Thirdly, it is proposed that the capacity to predict and to control the geomechanical performance of the host rock mass in which mining proceeds can prove or enhance the economic performance of the mine. This is proven in practice by the efficiency of resource recovery, measured in terms of volume extraction ratio, mine productivity or direct economic profitability.

These are simple practices, however the application of advanced methods of geoengineering mechanics in mine design and mine excavation operations is developing.

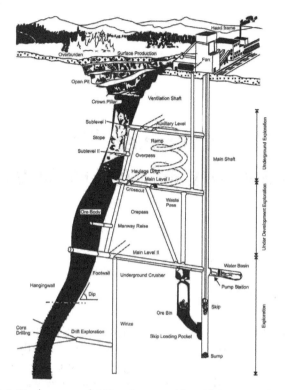

31.1 Underground Mine Layout, identifying openings, working places, and stages of the mine, as a shallow open pit mine transitions to a deep underground mine.

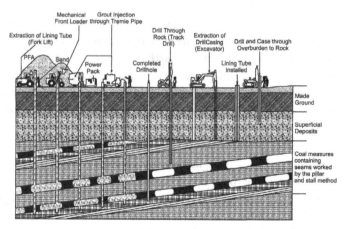

31.2 Sequence of Grouting Operations to Fill Large Voids Left in Coal Mines

Chapter 32
EARTHQUAKES, VOLCANOES AND TSUNAMIS

EARTHQUAKES

More than a million times a year, the Earth's crust suddenly shakes during an earthquake. Most of the world's earthquakes are fairly slight. A mild earthquake can feel like a truck passing. A severe earthquake can destroy roads and buildings and cause the sea to rise in huge waves. Earthquakes often happen near volcanoes and young mountain ranges and at the edges of the Earth's tectonic plates.

a) Earthquakes are caused by vibrations which occur when tectonic plates move against one another.

b) There are two types of waves caused by earthquakes: Surface waves, which are seismic waves that travel along the surface layer of the Earth, and body waves, which travel through the Earth's interior.

c) These waves are subdivided into P (Primary) waves, and S (Secondary) waves. P waves produce pressure by alternately expanding and compressing the materials through which they pass. S waves cause shaking due to oscillations that are perpendicular to the direction of propagation.

i) The standard mnemonic is P waves = Pressure; S waves = Shaking

d) The epicenter of an earthquake is the surface location that is directly above the center, the focus of the earthquake itself.

e) The Richter Scale is used to measure and annotate the energy of an earthquake. Each unit of increment of the scale indicates a 10-fold increase in a quake's power. For example, a 7.6 earthquake is 10 times more powerful than a 6.6 earthquake.

Seismology is the scientific study of earthquakes. Earthquakes generate several kinds of waves. As previously mentioned, these are classified in two categories: surface waves and body waves. Surface waves, are confined to the surface and shallow subsurface of Earth like ocean waves are confined to the surface and near surface of the water. They are very slow. There are two kinds of surface waves, named Love waves and Rayleigh waves after their discoverers. Love waves move like a snake. They exhibit horizontal shear, meaning a side-to-side or transverse shaking motion as they go forward. They cause buildings and other structures to fall over sideways. Rayleigh waves make the surface of Earth move like an ocean wave. They involve an elliptical motion of the ground that is actually retrograde, or opposite to the direction of travel. In other words, for a wave moving to the right, the movement of the rock will be counterclockwise. Body waves are also called p-waves and s-waves, named for the arrival, and P is the primary (P) or first arrival, and S is the secondary arrival. P-waves are the fastest of the seismic waves. They move by compressive and subsequent rarefaction of the rock media through which they move, in much the same manner as sound waves traveling through the air or vibrations from a hammer strike on a rock. S waves move in a transverse motion like the wave generated by whipping a jump rope. They have a similar motion to a Love wave and are slower than P waves.

Factors Affecting Destruction from Earthquakes

The extent of destruction from an earthquake depends on several factors:

Magnitude

Earthquake magnitude refers to how much energy the earthquake released. Magnitude is not necessarily proportional to damage. Under some conditions, such as high population density in the affected area, a relatively minor earthquake may do far more damage than a more powerful quake in a sparsely populated area.

There are two scales that are most commonly used to measure earthquake magnitude, namely the RICHTER scale and the MERCALLI scale. The Richter scale is a logarithmic scale (expressed in powers of 10) that reflects the amplitude of ground movement. Several steps are involved in establishing an earthquake's magnitude on the Richter scale. First, the location of the quake must be determined. It is done by noting the time a wave passed two widely separated locations and inserting the data into a set of equations to determine how far each station was from the origin of the quake. The next step is to estimate the intensity of ground motion at the earthquakes origin by using a correction factor that introduces some error into the calculation of magnitude. Measuring magnitude on the Richter scale does not yield precise results, but within certain limitations, it does provide a useful estimate of the strength of the earthquake. The Richter scale starts at zero and is open ended. An earthquake of magnitude 1 on the Richter scale probably would go unnoticed by anyone but an observer with a seismograph. Magnitude 4 earthquakes are mild but still perceptible, and earthquakes of magnitude 8 and above are devastating if they strike densely populated areas. The San Francisco Earthquake of 1906 is estimated to have been about 8.2 on the Richter scale, and two other earthquakes of the 20th century – one beneath the Pacific Ocean near Japan in 1933 and the other off the coast of Ecuador in 1953 – are thought to have ranked around 8.6.

The second common scale for measuring earthquake magnitude is the Mercalli scale. It uses Roman numerals to avoid confusion with the Richter scale and is based on easily observable effects, such as swaying of buildings.

Seismic moment is another measure of the size of an earthquake. Seismic moment is the product of the average amount of slip, the area of the rupture and the shear modulus, or strength, of the rocks affected. An expression of magnitude related to seismic moment is moment magnitude, which is approximately the same as Richter magnitude in range 3-7.

Duration

Some earthquakes last only a few seconds, but others have been known to last for a minute or two. As a rule, damage increases with duration.

Local Geology

The character of the subgrade greatly affects the extent of damage from an earthquake. The safest material on which to build, is solid rock, which transmits vibrations from an earthquake but maintains its structural integrity. More dangerous is unconsolidated sediment, which may exhibit liquefaction and extensive consequent damage where groundwater rises close to the surface. Fractured rock, as found in western North America, may absorb energy from earthquake waves and thus reduce the geographical extent of damage from a major earthquake. By contrast, in areas where geological formations stretch unbroken for hundreds of miles, as in portions of eastern North America, vibrations may travel great distances with much of their initial destructive potential intact. The 1906 San Francisco earthquake, for example, was barely felt at all outside California, but the New Madrid, Missouri, earthquakes approximately a century earlier were perceived over much of the United States east of the Mississippi River. It is ironic that the same processes that predispose a part of Earth's crust to major earthquakes may also confine damage from earthquakes through extensive fracturing, whereas comparatively earthquake free regions may have much greater potential for destruction because of the relatively undisturbed, contiguous rock that underlies them.

Time of Day

This is an important factor in determining the number of casualties from an earthquake. Deaths and injuries from collapsing buildings in rural areas, for example, are likely to be more numerous if an earthquake occurs at night or in the early morning when the population is indoors sleeping or having breakfast. In cities deaths and injuries may be more likely during the working day, when large numbers of pedestrians are exposed to harm from falling walls and decorative masonry.

Building Damages

Wood frame buildings have a reputation for standing up well to earthquakes because wood can bend and sway under earthquake vibrations, thus improving a building's chance to survive. Buildings of brick and stone construction, by contrast, are more likely to crumble and collapse because they lack wood's resiliency, although careful design and construction may increase the survivability of a non wooden structure. Possibly the worst building material in regard to earthquake resistance is adobe or clay brick, which was responsible for numerous fatalities during earthquakes in the early years of Euro-American settlement of the western United States.

Determining the potential for future rupture along a fault can be difficult because faults differ in their behaviours. One fault may show no activity for many centuries and then suddenly move 10 m (30 feet) or more, whereas another fault may exhibit virtually continuous, gradual movement and many small earthquakes. This variability in behaviour means that geologist cannot make accurate predictions of a faults future activity merely by establishing that the fault is active. Two selected faults, though both active, may move at greatly different intervals and exhibit equally great differences in rates of slip.

Another source of uncertainty in determining slip rate and history of activity is difficulty in estimating the ages of offset deposits or other features along a fault. In the Los Angeles area, for example, such estimates are unreliable in many or most locations where faults are active. Yet another problem in determining the true slip rate for a fault is the partial nature of information on components of slip. Data may be restricted to either the vertical or horizontal component. Such limited data may yield an unreliable estimate of slip along a fault, although the horizontal or vertical component alone may prove useful if the ratio of vertical to horizontal motion along a fault is known already. In situations involving a strike slip fault, where motion along a fault is primarily horizontal, the horizontal component alone may provide a reasonably accurate measure of the slip rate. This is the case along the San Andreas fault in California. Vertical component data reportedly have provided an approximation of true slip rates along the reverse slip faults in California's transverse ranges. The best information that can be supplied, in many cases, is an average figure for slip rate over a long period of time. As a rule, however, the higher the average slip rate, the more active the fault and the more closely it bears monitoring as a potential source of future earthquakes. In a few areas boundaries between major plates of the Earth's crust, such as along the San Andreas Fault show very high average slip rates, along faults perhaps 25 mm (one inch) or more per year. Active faults in other areas generally show less slip.

The future behaviour of a fault may be inferred from evidence including the rate of slip, the size of earthquakes and intervals between them and the amount of slip in each incidence of movement. No one set of criteria exists for determining how active a fault may be in the future.

Complicating the analysis is the fact that some active faults do not extend to the surface of the Earth and, on the surface, may display only faint evidence of recurring seismic activity. A highly damaging earthquake that struck Coalinga, California, in 1983 involved such a blind fault. Simply establishing that a fault is active does not allow geologists to make accurate predictions of its future activity. Ongoing measurements of seismic activity are useful tools for estimating the likelihood of earthquakes along a given fault in the future. Seismic data may be misleading, because ongoing earthquake activity, or the absence of it, along a fault does not necessarily reflect the potential of that fault for generating destructive earthquakes. An active fault may have very little potential to cause destructive earthquakes because it gradually releases its energy through creep, without actually generating earthquakes. Also, data on seismic activity along a particular fault may span too short a time to provide a useful means of predicting the behaviour of earthquake faults over a long period. Even when historical records are more extensive, as in parts of Asia, great uncertainties remain in reconstructing the seismic history of a given locality or region.

Despite these limitations, certain methods exist for estimating the size and frequency of possible future earthquakes along a fault. These methods involve, but are not limited to, analyzing the earthquake history of a region to find the major seismic event linked to a given fault and comparing a given fault's history of earthquake activity with that of other faults similar in structure and tectonic characteristics. Another approach is to use empirically established relations between earthquake magnitude and length of faults. It is assumed that the greater the dimensions of the fault surface involved in an earthquake, the greater magnitude the earthquake will have. One widely used method for estimating the most powerful earthquake that is likely to occur on a given fault rests on the assumption that half the total length of the fault may rupture in a particular earthquake. This approach is not fully reliable, however, because experience has shown that a major earthquake may involve rupture of anywhere from only a small percentage to almost the whole extent of a fault surface that existed prior to the earthquake. Another drawback to this method is difficulty in measuring the length of a given fault accurately. Much of a particular fault may lie hidden under sediment and water. It has been suggested that the whole San Andreas Fault might rupture in southern California, producing a single gigantic earthquake, but there are uncertainties about whether or not the geology of the region would allow such a single catastrophic event.

In summary earthquakes are caused when relative movement of plates or fault blocks overcomes shear resistance of a fault. Movement builds up elastic strain in rocks. The fault rupture and rock rebound release strain energy as ground shock waves. Most earthquakes originate at focus points less than 20 km deep. Surface displacement may be a few metres or absent. Fault breaks may extend over lengths of 1 to 100 km.

Size and Scale of Earthquakes

Ground movement is measured in different planes on seismographs.

Magnitude defines the size of an earthquake on the Richter scale: $\log_{10}$ of the maximum wave amplitude in microns on a Wood Anderson seismograph 100 km from the epicenter (point on the surface above the focus).

Intensity is the scale of earthquake damage at any one point, described on the modified Mercalli scale, and declining away from the epicenter.

Damage relates largely to peak ground acceleration, also to peak velocity, frequency, and duration.

Peak horizontal acceleration up to > 1.0 g at intensity X, but < 0.1g at intensity Y. Vertical acceleration is usually about half of these values.

Duration usually < 10 s for magnitude 5, may last 40 s for magnitude 8; increases away from epicenter.

Earthquake Prediction

Most are on plate boundaries and 90% are on subduction zones.

Some occur on interplate faults. Britain has up to M5, and the Mississippi Valley earthquakes of 1811 reached M7.8. Also due to magma movement under volcanoes.

Prediction is based on side effects during strain accumulation, but monitoring foreshocks, uplift, dilation, seismic velocities, gas emissions and groundwater levels has revealed no patterns for reliable predictions.

Historical data may indicate fault traces with no recent movements, or seismic gaps, where a future earthquake is more likely.

Some faults slip smoothly. The Cienega Winery in California has its foundations displaced 15 mm/year by the San Andreas Fault, but no earthquake damage.

Increased water pressures reduce shear strength and cause fault movement before large strain energy accumulation. Pumping water into deep wells causes premature small quakes, but legal complications make serious earthquake control impossible.

Construction in Seismic Zones

Adobe and dry stone walls fail under horizontal acceleration of 0.1 g, but good low rise timber buildings can withstand most earthquakes.

Reinforced concrete structures need bracing to stop rhombohedral collapse. This can be provided by massive, resistant shear walls, or diagonal steelwork.

Rebars must be integrated across the intersection of columns/beams/walls/slabs.

Pile cap failures are restrained by tie beams and integrated basement structures.

Buildings and bridges can be isolated on rubber spring blocks, and steel springs can act as energy absorbers to stabilize structures.

Precautionary provisions add 5-10% to construction costs. Later modifications are more expensive.

It is important to avoid decorative appendages which can fall off in active earthquake zones.

Use land zoning to avoid areas of deep soft soils and known fault traces. Any displacement of Holocene soils indicates modern activity on a fault.

New building in California is prohibited within 15 m of active faults. Wider zones apply to larger buildings and less well mapped faults.

Deep Soils and Earthquakes

Soft soils do not dampen ground vibrations. They amplify them. Buildings on soft soil suffer much worse earthquake damage than those on bedrock.

Wave amplitude may double passing from rock to soil. The dominant natural period of the shock waves also increases, from about 0.3 seconds in solid rock, to 1-4 seconds on soils. The natural period further increases with soil depth, and with distance from the epicenter.

Buildings have a natural period of about N/10 seconds (N=number of stories). Maximum damage is due to resonance, when periods of building and soil match.

Deep soft soils have long periods which match those of high rise buildings susceptible to more catastrophic damage as in the Mexico City earthquake in 1985.

Compared to adjacent bedrock, soft soils cause damage 1 to 3 intensities higher.

Earthquakes and Liquefaction

An earthquake shakes the ground back and forth like noodles in a frying pan. The movement is a vivid demonstration of Newton's Second Law, $f = ma$, where f is force, m is mass, and a is acceleration provided by the earthquake. The inertia of the soil mass contributes to horizontal body forces, and can trigger multiple landslides, particularly in marginally stable ground.

Soil liquefaction is a highly destructive phenomenon when shaking of a saturated, low-density granular soil breaks down its structure so that it collapses in on the soil water. Intergranular contact stress is transferred to pore water pressure so the soil becomes a dense liquid the same as quicksand. After a granular soil has liquefied and shaking stops, soil grains suspended in water settle out so excess pore water rises to the ground surface, carrying along sand to make small sand volcanoes. The existence of ejected sand in layered sediments allows geologic dating of earlier earthquakes in order to determine their periodicity.

VOLCANOES

A volcano is defined as an opening in the Earth's crust through which magma escapes to the surface where it is transformed into lava. More specifically, the word volcano refers to mountains produced by volcanic activity known as volcanism or volcanism. The words volcano, volcanism, and volcanism all are derived from the Latin Volasnus, or Vulcan, the god of fire in Roman mythology.

Most volcanoes are found near the coast or under the ocean. They usually form at plate edges. Here crust movement allows hot molten rock called magma to rise up from inside the Earth and burst through the crust. Hot magma is called lava when it flows out of a volcano. Ash, steam, and gas spew out and cause great destruction.

a) Volcanoes often form along oceanic ridges.
b) There are three basic types of volcanic structures:
    i) Shield volcanoes are broad and gently sloping. They are built from fluid basaltic lava.
    ii) Cinder cones are small volcanoes built of pyroclastics, igneous rock texture resulting from individual rock fragments that consolidate and are ejected during an eruption.
    iii) Composite cones are volcanoes formed from a combination of both lava flow and pyroclastic materials.

c) The nature of a volcanic eruption depends on factors such as composition and temperature of lava.
g) The formation of magma, lava within the earth is determined by pressure, temperature, and constitution.
h) When volcanoes erupt they eject lava, gases, pulverized rock and glass.
f) Volcanic regions may also contain:
    i) Volcanic necks: steep sided erosional remnants of lava that at one time occupied a volcano vent
    ii) Craters: depressions at the summit of a volcano
    iii) Fissure eruptions: lava is extruded from narrow cracks (fissures) in the crust
    iv) Calderas: large depressions caused by the collapse or ejection of a volcano's summit area.

In a volcanic eruption, magma or the hot gases from magma escape from an underground reservoir to the surface through a relatively narrow vent, or conduit. Eruptions differ greatly in character from one volcano to another and sometimes with the history of the same volcano. Some eruptions are extremely violent and involve great outbursts of ash, gas and lava. These eruptions produce cinder cones and composite volcano.

A cinder cone is made up of fragments of rock ejected from the vent. Paricutin in Mexico is one example of a cinder cone. These fragments tend to be low density rock that are formed when dissolved gases in the magma bubble out of solution as the rock solidifies, producing pumice that appears to be rock foam full of small cavities that give the rock a frothy texture. Rock fragments between about 12 to 50 mm in diameter are called cinders and constitute most of the cinder cone. Cinders are distinct from ash. It is finer material that winds may carry for great distances away from the volcano. Although there is a significant component of ash to every cinder cone. Mixed in with the cinders, in most cases, are volcanic bombs that are formed as large masses of lava are rejected intact. Cinder cones may grow rapidly, rising hundreds of feet in the first few days or weeks of their existence. Cinder cones are typically steep-sided. The slope of a newly formed cinder cone can be 28°. The crater tends to be large and to have a rim higher on one side than the other because of prevailing winds that carry the volcano's output in a given direction. Cinder cones may occur virtually anywhere the appropriate kind of magma rich in dissolved gases can reach the surface. Clusters on cinder cones are commonplace. Eruptions of cinder cones may include lava flows.

A composite volcano is more complex than a single cinder cone. Composite volcanoes are made up of layers of cinder and ash alternating with lava. Because of these alternating strata, composite volcanoes are known as stratovolcanoes. A stratovolcano has steeply sloping sides, as cinder cones have, but has greater structural strength due to the rigid lava layers inside it. Stratovolcanoes may reach thousands of feet in height. Examples of straovolcanoes are Vesuvius in Italy, Mount Fuji in Japan, and Mount Saint Helens in the United States. Most stratovolcanoes are concentrated in two parts of the world; in the ring of fire, a belt of intense volcanic and earthquake activity encircling the Pacific Ocean basin's and in the Mediterranean Sea. Eruptions of stratovolcanoes involve release of hot gases, ash, ciders, bombs, and lava.

Basaltic Volcanoes lie on divergent plate boundaries (Iceland), or on plates away from boundary disturbances (Hawaii), where magma is generated from mantle plumes. They produce large flows of mobile lava in quiet, effusive eruptions, with only limited fountaining or explosions.

These volcanoes are tourist attractions, which may threaten fixed structures, but offer minimal threat to life.

Prediction of Eruptions is largely based on volcanic inflation (uplift) and seismic monitoring, with successful forecasts of repetitive basalt emissions. Scale and size of explosive eruptions cannot be readily predicted, nor can their precise timing and location within the volcanic area.

Explosive Volcanoes occur on convergent plate boundaries (e.g. Krakatoa, St. Helens), as magna is generated by subduction melting. They have viscous magma, andesite or rhyolite, so gas pressures can build up. They produce ash clouds, explosive blasts and very dangerous pyroclastic flows (of hot gas and ash) which turn into lahars (mud flows) lower down valleys. Lava flows are minor and short.

Explosive eruptions are dangerous and uncontrollable, and must be avoided.

TSUNAMIS

Tsunamis are tidal waves generated by a hurricane or, by an underwater earthquake. An earthquake near to the coast can start a wave motion at sea. In the ocean the wave is low, but as it nears the shore the front of it slows and water behind builds up to form a huge tsunami.

Tsunamis are a special type of oceanic surface gravity waves. Formally, they belong in the same fundamental classification as ordinary sea waves that can be observed daily at the beach. Tsunamis, however, are distinct in their mode of generation, in their characteristic period, and in their effect upon the shore where they impinge. Unlike ordinary waves generated by surface winds, tsunamis are often produced by a shift in the position of the seafloor. Although ocean floor shifts can originate from undersea landslides and volcanic eruptions, the most common cause is submarine earthquakes. Reflecting this, tsunamis are sometimes called seismic sea waves. Compared with ordinary wind driven waves, seismic sea waves have a much longer period and wavelength and, as a result, have a profoundly different effect on coastlines. Shoreline defenses such as riprap and breakwaters are designed to withstand storm waves with period of 6-10 s and wavelengths of 100 m or so. With periods of 200-200s and wavelengths of tens of kilometres, tsunamis overwhelm most defenses.

Tsunamis produce up thoughts of killer waves. Indeed, the amplitude of sea waves associated with the greatest earthquakes can be impressive. Open ocean heights of 5 m to 10 m are possible. Upon reaching shore, the waves shoal and are amplified by a factor of 2 or 3. Tsunamis over a metre or two in height are actually rare, needing about a magnitude 8 earthquake for their production. On a global average less than one magnitude 8 earthquake occurs per year; and of these, perhaps one in ten is located under the ocean and capable of sea wave excitation. In earthquake studies, the largest damaging event receive most of the publicity, but it is the smaller events that are useful scientifically, simply because they are much more frequent. A tsunami of a few centimetres height is easily observable by modern pressure sensors even in the open sea. Waves of this size can be generated by earthquakes of magnitude 6.5, and occur several times per year.

The life of tsunami waves covers three phrases: generation, propagation, and disposition at the receiver.

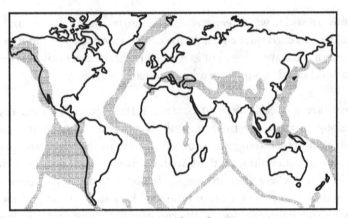

32.1 Active Earthquake Zones

Chapter 33
GLOBAL WARMING, CLIMATE CHANGE AND SOIL DEHYDRATION

Global Warming

Climatologists state the world is warming. Climatic zones are shifting. Glaciers are melting. Sea level is rising. These are not hypothetical events; these changes are taking place, and scientists expect them to accelerate over the next years as the amounts of carbon dioxide, methane, and other trace gases accumulating in the atmosphere increase due to human activities.

Global warming is one of the most controversial earth science issues of the 21st century, challenging the structure of our global society. The problem is that global warming is not just a scientific and engineering concern, but encompasses economics, sociology, geopolitics, local politics, and an individuals' choice of lifestyle. Global warming is caused by the massive increase of greenhouse gases, such as carbon dioxide, in the atmosphere, resulting from the burning of fossil fuels and deforestation. There is clear evidence that we have elevated concentrations of atmospheric carbon dioxide to their highest level for the last half million years and maybe longer. Scientists believe that this is causing the Earth to warm faster than at any other time during the past one thousand years. The Intergovernmental Panel on Climate Change (IPCC), recently declared that the scientific uncertainties of global warming are essentially resolved. They report that there is clear evidence for a 0.6°C rise in global temperatures and 20cm rise in sea level during the 21st century. The IPCC predicts that global temperatures could rise by between

1.4ºC and 5.8ºC and the sea level could rise by between 20cm and 88cm by the year 2100. In addition, weather patterns will become less predictable and the occurrence of extreme climate events, such as storms, floods, and drought, will increase.

Earth's Natural Greenhouse

The temperature of the Earth is controlled by the balance between the input from energy of the sun and the loss of this back into space. Certain atmospheric gases are critical to this temperature balance and are known as greenhouse gases. The energy received from the sun is in the form of short wave radiation. On average, about one-third of this solar radiation that hits the Earth is reflected back to space. Of the remainder, some is absorbed by the atmosphere, but most is absorbed by the land and oceans. The Earth's surface becomes warm and as a result emits long wave infrared radiation. The greenhouse gases trap and re-emit some of this long wave radiation, and warm the atmosphere. Naturally occurring greenhouse gases include water vapour, carbon dioxide, ozone, methane, and nitrous oxide, and together they create a natural greenhouse or blanket effect, warming the Earth. Despite the greenhouse gases often being depicted as one layer, this is only to demonstrate their blanket effect, as they are in fact mixed throughout the atmosphere.

According to evidence from ice cores, atmospheric CO_2 levels during the ice age were about 200 ppmv compared to pre-industrial levels of 280 ppmv – an increase of over 160 billion tonnes – almost the same CO_2 pollution that we have put into the atmosphere over the last 100 years. This carbon dioxide increase was accompanied by a global warming of 6ºC as the world freed itself from the grips of the last ice age. Although the ultimate cause of the end of the last ice age was changes in the Earth's orbit around the sun, scientists studying past climates have realized the central role atmospheric carbon dioxide has recently played. Field surveys demonstrate that the level of pollution that we have caused in one century of industrialization is comparable to the natural variations which took tens of thousands of years.

Pollution Sources

Carbon dioxide emissions are not evenly produced by countries. The first major source of carbon dioxide is the burning of fossil fuels, since a significant part of the carbon dioxide emissions comes from energy production, industrial processes, and transport. These are not evenly distributed around the world because of the unequal distribution of industry, therefore, any global agreement would affect certain countries' economies more than others. Currently the industrialized countries should bear the main responsibility for reducing emissions of carbon dioxide to about 22 billion tonnes of carbon per year. North America, Europe, and Asia emit over 90% of the global industrially produced carbon dioxide.

Eliminating all the risks of climate change is impossible because carbon dioxide emissions, the chief human contribution to global warming, are unlike conventional air pollutants, which stay in the atmosphere for only hours or days. Once carbon dioxide enters the atmosphere, much of it remains for over a hundred years. Emissions from anywhere on the planet contribute to the global problem, and once headed in the wrong direction, the climate system is slow to respond to attempts at reversal. As with a sink that has a large faucet and a small drain, the practical way to lower the level is by dramatically cutting the inflow. Holding global warming steady at its current rate would require a worldwide 60 to 80 percent cut in emissions, and it would take decades for the atmospheric concentration of carbon dioxide to stabilize.

The second major source of carbon dioxide emissions is a result of land use changes. These emissions come primarily from the cutting down of forests for the purposes of agriculture, urbanization, or roads. When large areas of rainforests are cut down, the land often turns into less productive grasslands with considerably less capacity for storing CO_2. Here the pattern of carbon dioxide emissions is different, with South America, Asia, and Africa being responsible for over 90% of present day land use change emissions, about 4 billion tonnes of carbon per year. North America and Europe were early to clear their own landscape by the beginning of the 20th century. In terms of the amount of carbon dioxide released, industrial processes still significantly outweigh land use changes.

Climate Change

Many scientists believe that the human induced or anthropogenic enhanced greenhouse effect will cause climate change in the near future. Natural climate change does occur on human timescales and we should be prepared to adapt to it. Climate change can manifest itself in a number of ways, for example changes in regional and global temperatures, changing rainfall patterns, expansion and contraction of ice sheets, and sea level variations. These regional and global climate changes are responses to external and/or internal forcing mechanisms. An example of an internal forcing mechanism is the variations in the carbon dioxide content of the atmosphere modulating the greenhouse effect, while a good example of an external forcing mechanism is the long term variations in the Earth's orbits around the sun.

Climate Change versus Global Warming

The current scientific consensus is that changes in greenhouse gas concentrations in the atmosphere do cause global temperature change. Predicting climate change is complex because it encompasses many different factors, which respond differently when the atmosphere warms up, including regional temperature changes, melting glaciers and ice sheets, relative sea level change, precipitation changes, storm intensity and tracks, El Niño, and ocean circulation. This linkage between global warming and climate change is complicated by the fact that each part of the global climate system has different response times. For example, the atmosphere can respond to external or internal changes within a day, but the deep ocean may take decades to respond, while vegetation can alter its structure within a few weeks but its evolutionary composition can take up to a century to change. There is the possibility of natural forcing which may be cyclic, for example, there is good evidence that sunspot cycles can affect climate on both a decadal and a century timescale. There is also evidence that since the beginning of our present interglacial period, the last 10,000 years, there have been climatic coolings every 1,500 ± 500 years, of which the Little Ice Age was the last. The Little Ice Age began in the 17th century and ended in the 18th century and was characterized by a fall of 0.5 - 1°C in Greenland temperatures, significant shift in the currents around Iceland, and a sea surface temperature fall of 4°C off the coast of West Africa, 2°C off the Bermuda Rise, and of course ice on the River Thames in London, all of which were due to natural climate change. It is important to differentiate natural climate variability from global warming. The different parts of the climate system interact, and they all have different response times.

Scientists are predicting that global warming could warm the planet by between 1.4 and 5.8°C in the next 100 years, causing very significant problems for humanity. Each year, the effects of climate change are coming into sharper focus. Ice sheets and glaciers are melting faster than expected, sea levels are rising more rapidly than ever in recorded history, plants are blooming earlier n the spring, water supplies and habitats are in danger, birds are being forced to find new migratory patterns.

Soil Dehydration

Geotechnical field studies in sensitive marine clays along the St. Lawrence River and its tributaries show that insitu soil moisture contents have decreased by and about 30% in the upper few metres in the past few decades, resulting in abnormal soil consolidation, settlement and cracking in buildings necessitating soil wetting approaches and revised design and construction methods not previously considered industry standards.

The history of the global warming hypothesis clearly shows that science and engineers are deeply influenced by society and vice versa. The essential science of global warming was carried out 50 years ago under the perceived necessity of geosciences during the Cold War, but it was not taken seriously. If global warming is causing increased soil dehydration, then building code revisions for foundation design, and the like will be required.

Adaptation and Mitigation

The sensible approach to preventing the unfavourable effects of global warming would be to cut carbon dioxide emissions. Scientists believe a cut of between 60 and 80% is required to avoid the worst effects of global warming. Petrochemical supporters argue that the cost of significant cuts in fossil fuel use would severely affect the global economy. The IPCC believes that we must adapt to climate change and thereby deal with soil dehydration. This is a controversial strategy.

Global Warming Solutions

Governments are slowly coordinating plans to reduce carbon dioxide emission, however, there are concerns over costs. There is increasing consideration for the alternatives to solving the problem of global warming. There are five main areas of remedial action:

a) CO_2 removal from industrial processes can contribute substantially to a reduction in atmospheric CO_2, provided that the methods are within the concepts of sustainable development.

b) Use less energy and thus produce less carbon dioxide.

c) There are renewable/alternative energy sources, i.e. energy sources which do not produce a net amount of carbon dioxide in the atmosphere.

Biomass is promising in the short term, which by the year 2020 could produce a third of the global energy. When the biomass is growing it absorbs carbon dioxide from the atmosphere which is only returned when it is burnt as fuel and thus there is no net increase in atmospheric carbon dioxide. Most promising from the long term is solar energy, while wind power is thought to be an intermediate solution, particularly in countries such as the UK, where sunlight cannot be guaranteed. Many countries are discussing renewing their nuclear programmes as a noncarbon emission energy source, but problems of safety and dumping nuclear waste remain the primary objections. Cars that run on fuel cells, hydrogen, and compressed air are developing.

d) The possibility exists of removing carbon dioxide from the atmosphere either by growing new forests or by stimulating the ocean to take up more.

e) Planetary GeoEngineering

The world's slow progress in cutting carbon dioxide emissions and the danger that the climate could take a sudden turn for the worse require policymakers to take a closer look at emergency strategies for curbing the effects of global warming. These strategies, sometimes called planetary geoengineering, but not in the geological or civil engineering sense, envision deploying systems on a planetary scale, such as launching reflective particles into the atmosphere or positioning sunshades to cool the earth. These strategies could cool the planet, but they would not stop the buildup of carbon dioxide or lessen all its harmful impacts. For this reason, geoengineering has been unacceptable to those committed to reducing emissions.

The term planetary geoengineering with respect to global warming, refers to a variety of strategies designed to cool the climate. For example, some would slowly remove carbon dioxide from the atmosphere, either by manipulating the biosphere such as by fertilizing the ocean with nutrients that would allow plankton to grow faster and thus absorb more carbon, or by directly scrubbing the air with devices that resemble large cooling towers. Increasing the earth's albedo offers another promising method for rapidly cooling the planet. Schemes that would alter the earth's albedo envision putting reflective particles into the upper atmosphere, much as volcanoes do already. Every few decades, volcanoes validate the theory that it is possible to engineer the climate. Such schemes offer rapid impact with relatively little effort. For example, just one kilogram of sulfur well placed in the stratosphere would roughly offset the warming effect of several hundred thousand kilograms of carbon dioxide. Other schemes include seeding bright reflective clouds by blowing seawater or other substances into the lower atmosphere. Substantial reductions of global warming are also possible to achieve by converting dark places that absorb lots of sunlight to lighter shades – for example, by replacing dark forests with more reflective grasslands. Engineered plants might be designed for the task.

There is general agreement the geoengineering strategies are low cost. The total expense of the most cost effective options would amount to perhaps as little as a few billion dollars, just one percent, or less, of the cost of dramatically cutting carbon dioxide emissions.

f) Cooling the planet through planetary geoengineering will not fix all of the problems related to climate change. Offsetting warming by reflecting more sunlight back into space will not stop the rising concentration of carbon dioxide in the atmosphere. Eventually, much of that carbon dioxide ends up in the oceans, where it forms carbonic acid. Ocean acidification is a catastrophe for marine ecosystems, for the 100 million people who depend on coral reefs for their livelihoods, and for the many more who depend on them for coastal protection from storms and for biological support of the greater ocean food source. Over the last century, the oceans have become more acidic, and current projections suggest that without a serious effort to control emissions, the concentration of carbon dioxide will be so high by the end of the century that many organisms that make shells will disappear and most coral reef ecosystems will deteriorate, reducing the marine fishing industry.

Conclusions

All of these remedial technologies are feasible and a combination of them could be used to combat global warming, although they each have drawbacks. Removal of carbon dioxide during industrial processes is challenging and costly, because not only does the CO_2 need to be removed, but it must be stored somewhere. Recovered CO_2 does not all need to be stored, some may be utilized in enhanced oil recovery, the food industry, chemical manufacturing producing soda ash, urea, and methanol, and the metal processing industries. CO_2 can also be applied to the production of construction material, solvents, cleaning compounds and packaging, and in waste water treatment. In practice, most of the carbon dioxide captured from industrial processes would have to be stored. It has been estimated that theoretically two thirds of the CO_2 formed from the combustion of the world's total oil and gas reserves could be stored in the corresponding reservoirs. Oceans could also be used to dispose of the carbon dioxide. The major problem with all of these methods of storage is safety. Carbon dioxide is a very dangerous gas because it is heavier than air and causes suffocation.

Consequently, a combination of improved energy efficiency and alternative energy is the better solution to global warming. From the safety and environmental perspective, the storage of carbon dioxide either underground and/or in the ocean is considered to be not feasible, however helpful this would be in the short term.

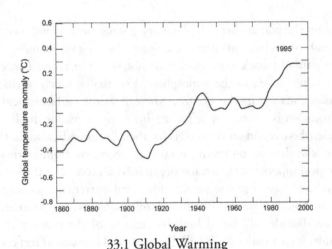

33.1 Global Warming
Changes in global average surface temperature (surface air temperature over land and sea surface temperatures combined) from 1860 to 1995 relative to the period 1961 to 1990

Chapter 34
GEOENVIRONMENTAL ENGINEERING

The objectives of geoenvironmental engineering include: (1) safe containment of wastes, (2) cleanup of contaminated ground and groundwater, (3) prevention of contamination and pollution, and (4) enhancement of the natural and man-made environment. The attainment of these goals inevitably involves activities on, in, or with the Earth. The properties and behaviour of soil, rock and groundwater over a range of conditions are of major importance in geoenvironmental engineering. It is important to identify the soil, rock and groundwater characteristics that are important in dealing with geoenvironmental problems. They include contaminant hydrogeology and contaminant migration, clay liners, geosynthetics in waste containment systems, landfill covers, insitu containment and treatment of contaminated soil and groundwater, and contaminated soil management.

For environmental risk assessment it is important to understand that when contaminants, receptors and pathways are present, there is a potential risk.

Hydrogeologic Environment

While advances into understanding the complexities of contaminant migration through porous media and fractured rock are progressing, it is important to understand contaminant migration and the design of remedial systems, within an accurate characterization of the hydrogeologic environment.

It is important to understand the hydrostratigraphy comprising the soil, rock and groundwater conditions, and other factors that control fluid flow at a site in order to understand or control contaminant migration.

Saturated and Unsaturated Zones

In the subsurface, the region above the water table is referred to as the vadose or unsaturated zone, whereas the region below the water table is referred to as the phreatic or saturated zone. However, the terms saturated and unsaturated zones require some clarification because the distinction between the two zones is actually based on fluid pressure and not degree of saturation. Typically, the water table is defined by the water level in a shallow well or piezometer screened across the vadose and phreatic boundary. The water pressure is less than atmospheric in the vadose zone, and greater than atmospheric in the phreatic zone. The capillary fringe is a zone above the water table where fluid saturations are high, close to 100%, but fluid pressure is low owing to tension of the sediment-surface-water contact. The capillary fringe is thicker in fine-grained sediments than coarse grained sediments because of the greater tension created by the smaller pore opening.

Aquifers and Aquitards

An aquifer is a saturated permeable geologic unit that can transmit significant quantities of water to wells under ordinary hydraulic gradients. Unconsolidated sands and gravels, sandstones, limestones and fractured rock are common aquifers. In contrast, an aquitard is a saturated low permeability geologic unit that transmits significantly less water than adjacent aquifers. Clays, shales and unfractured crystalline rock are common aquitards.

Landfills

Typically a waste landfill requires barrier liner systems to minimize the outward flux of liquids such as perched leachate, or gases such as methane generated from municipal solid waste. The primary objective is to limit the physical escape of liquid to either surface water or groundwater. A secondary objective is to limit chemical migration by the process of diffusion.

Contaminant Identification

Contaminant identification within geoenvironmental problem identification and risk management is concerned principally with establishing the source, abundance, distribution and direction, and speed of movement of contaminants in the subsurface media. Sometimes we simplify this to a generator — pathway — receptor model.

Typical Contaminants

In general, contaminants are classified according to their chemical characteristics. There are two main types of chemical contaminant: inorganic and organic.

Inorganic Contamination

This is usually associated with the presence of metals, but also includes cyanides and other anions such as chlorides, sulphates and nitrates. Metal contamination is common at former industrial sites, and can be associated with activities such as mining, smelting, steel production, landfarming, scrap metal yards, vehicle maintenance, manufacturing, sewage treatment and the like. In particular, arsenic, cadmium and lead are known to be acutely toxic to both plants and humans, whereas zinc, copper and nickel are primarily phytotoxic (harmful to plants).

Cyanides are associated with mining, manufactured gas plants and waste disposal facilities, whereas sulphates are produced by numerous industrial processes. Both represent a hazard to human and ecological receptors, in addition to acting as corrosive agents on building foundation.

Examples of Industries/Activities and their Contaminants

Inorganic Contaminants	Organic Contaminants

Examples
    Iron (Fe), manganese (Mn) - usually of concern for aesthetic reasons in water supplies
    Heavy Metals – silver (Ag), mercury (Hg), cadmium (Cd), Zinc (Zn), copper (Cu), lead (Pb), chromium (Cr), arsenic (As)
    Acids and bases
    Cyanide
 	
Typical Sources
    Mining and smelting operations
    Mine tailings and tailings ponds
    Landfills
    Sewage
    Metal plating operations (Ag, Cr)
    Paint, painting operations (Pb, Cd)
    Pulp mills (Hg)
    Wood treatment (Cu, Cr, As)
    Auto wrecking yards
    Industrial and municipal landfills
    Thermal power stations
    Imported fill soils (when imported to a site from one of the primary sources areas)
	Examples
	    Polynuclear aromatic hydrocarbons (PAHs)
	    Phenols
	    Pesticides
	    Nonchlorinated solvents
	    Chlorinated solvents
	    Polychlorinated biphenyls
	    Petroleum hydrocarbons
	
	Typical Sources
	    Fuel dispensing
	    Electrical conductors
	    Bulk storage depots
	    Sewage
	    Metal plating operations
	    Paint, painting operations
	    Wood treatment
	    Auto industry
	    Home heating
	    Industrial and municipal landfills
	    Power stations

Organic Contamination

Organic contaminants include petroleum products (from gasolines to heavy fuel oils to bitumen), coal tar, solvents (including chlorinated compounds), phenols, dioxins and expanded polychlorinated biphenyls (PCBs). These contaminants differ widely in their chemical composition and inherent toxicity. In considering the risk associated with these contaminants, it is important to recognize that organic contaminants are usually mixtures of many individual chemical compounds (gasoline, for example, comprises several thousand of compounds), each exhibiting different chemical, physical, biological and toxicological properties, and that this can result in multiphase environmental partitioning following a spill event. Many organic contaminants, most notably benzene and polynuclear aromatic hydrocarbons (PAHs), are considered to be carcinogenic to humans.

Common vapor phase contaminants include methane, carbon dioxide, carbon monoxide, hydrogen sulphide and sulphur dioxide, as well as low molecular weight volatile organic compounds (VOCs), which exist either in vapor or liquid phases. This substance presents a variety of risk, including combustion, asphyxiation, corrosion and health impacts following inhalation.

Environmental Risk Assessment

The issue of contaminated land carries with it significant environmental and economical implications. A key question is how best to fulfill the interests of all those with a stake in contaminated land (owners, potential vendors/ purchases, consultants, the public, the regulators, and more) while ensuring that the land is returned to a condition that is protective of the users and does not present a long term environmental liability.

Increasingly, in lieu of full depth site remediation, contaminated sites are being managed through consideration of the risks they pose to human and environmental health. The process of identifying and evaluating the significance of risks is known as risk assessment. The process through which identified risks are controlled or mitigated is known as risk management.

Environmental risk assessment is a framework for evaluating the likelihood of adverse effects on humans and the wider ecological environmental impact resulting from exposure to a contaminant. Use of risk assessment at contaminated sites provides a customized, clearer, more comprehensive means of discriminating between, and identifying appropriate solutions for the potential issues of concern associated with a contaminated site than by direct comparison with generic nonhealth-based standards.

Typically, the risk based approach is often most valuable in situations where financial or technical constraints rule out a conventional solution, for example, where contaminated portions of the site underlie existing building structures or services, where landfill disposal is prohibitively expensive, or where the contaminated material is deep lying and cannot be removed. However, its application is by no means confined to such situations, and risk based site management is frequently applied in a wider sense as a recognized cleanup framework in many jurisdictions across Canada, the USA and Europe.

Environmental risk assessment at contaminated sites entails characterization of the source of environmental hazard, identification of exposure routes through which a contaminant may come into contact with a receptor, assessment of the relationship between the contaminants and the adverse effects produced, if any, and, finally, estimation of the effect (CCME 1996). The primary role of a risk assessment is a technical framework in which potentially contaminated land can be examined, target remediation levels derived, and the environmental issues caused by the presence of contamination managed to protect human and environmental health.

During risk assessment, a major requirement is to identify the key contaminants, pathways and receptors associated with a site, and, at the same time, rule out possible risk scenarios found to be of negligible or no significance.

Environmental impact assessment is often the final stage of the risk assessment process. It combines the information from the exposure analysis and the toxicity analysis to estimate the magnitude and nature of the risk, as a probability, if any, to which the site receptors may be subjected.

Risk characterization comprises two parts; firstly, estimating risks by comparing estimated exposure with the acceptable exposure limit, and secondly, a description of the estimated risks with respect to site specifications, background conditions, future landuse, uncertaintities, and other technical, economic and societal factors.

The risk characterization from the environmental risk assessment (ERA) and the environmental impact assessment (EIA) form the basis for subsequent environmental risk management decisions to arrive at an exposure scenario which is considered to be the most protective of human and ecological health.

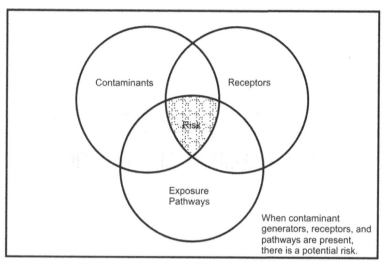

When contaminant generators, receptors, and pathways are present, there is a potential risk.

34.1 Environmental Risk Assessment Components

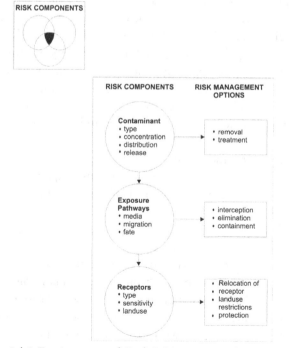

34.2 Environmental Risk Management Options

Chapter 35

GEOENVIRONMENTAL SITE ASSESSMENT

General

Since the 1970's the public has become more interested in the effects of large scale projects involving municipal development, highways, airports, water and sewage treatment plants, pipelines, river diversions, industrial and petrochemical plants, mining and forestry facilities, landfills and spills. In many instances the public has objected to property rights. There is growing public support and enforceable legislation in developing jurisdictions for the environmental assessment; and particularly a geoenvironmental site assessment with subsurface sampling and testing is an essential component of all stages of project planning, implementation, operation, and site decommissioning.

Typically an Environmental Assessment Act (EA Act) is applicable to government ministries and agencies and to municipalities. Private sector projects may be designed with a government on a case by case basis.

Purpose

An EA Act is intended to provide for the protection, conservation and wise management of the environment through sound planning and informed decision making. Such legislation requires those proposing a project, the proponent, to submit an Environmental Assessment of the undertaking to the regulatory body and/or the Minister of the Environment.

Systematic evaluation of alternatives and net effects establishes that the project is a preferred solution to a problem and ensures minimal environmental degradation. While mitigation techniques are likely to be required, avoidance of environmental impacts is the aim. Reduced costs to the public and to the environment are the benefits of this proactive versus reactive approach.

Environmental Assessment Types

There are two types of environmental assessments which can be prepared, a class or an individual EA.

An approved Class EA provides a series of planning requirements, for use solely with that particular group (class) of projects. An undertaking can be elevated to an individual EA for such reasons as discovery of unanticipated environmental effects of significant public concern.

An individual environmental assessment (including the Class EA parent document) may include the following:

- a description of the alternatives that were considered;
- their net environmental effects;
- advantages and disadvantages;
- an explanation of why the project was chosen;
- detailed description of the project for which approval is requested;
- net effects of the project;
- project design;
- site specific location for the project;
- preliminary construction schedule.

Environmental Control Procedures

A legal requirement of the approval to proceed issued by the Ministry of the Environment (MOE) or the Environmental Assessment Board is that the proponent is bound to carry out the mitigation techniques in accordance with what has been outlined in the environmental assessment document and the conditions of approval. This means that these techniques must be outlined in the tender documents and the plan and profile drawings by the engineering consultants so that every contractor bidding on the work is aware it represents part of the bid package.

This environmental site inspector's role is to ensure that the methods for mitigating or avoiding the environmental impacts identified in the contract documents are put into practice, plus be aware of any statutory or permit requirements, terms and conditions, and other requirements of the project approval.

Mitigation practice need not be complicated, more often it is common sense.

Construction Equipment Maintenance

The inspector should take precautions to prevent and rapidly cleanup the spillage of oils, fuels, and other waste solutions generated during equipment maintenance. In the event of a spill, contact the MOE Spills Action Centre.

Refuelling and maintenance of equipment cannot be allowed in or adjacent to watercourses but must be performed at remote storage areas.

Spilling Areas

Normally structures and pipe alignments are selected to give the foot of steep slopes and valley walls, which are prone to ground movement and slippage, as wide clearance as possible. These would include what areas require stabilization and what construction material is considered suitable for that application.

Right-of-Way

The construction right-of-way is designed to minimize areas of disturbance. In the vicinity of areas designated as critical such as stream crossings, mature tree stands, etc. The right-of-way must be limited in order to reduce adverse effects. Such limitations may necessitate a change in excavation procedures near these critical areas.

Clearing of Vegetation

Prior to any clearing of vegetation, the routing should be examined to determine if it is necessary to remove all vegetation along the construction right-of-way. All trees need not be removed if the pruning of low branches will suffice. Restoration of vegetation, including interim measures, should be done in stages as soon as possible after clearing. Trees that are removed are usually replaced on a one-for-one basis.

Stockpiling of Excavated Material

There are important procedures for stockpiling topsoil, excess excavated material, bouldery remnants, etc. Methods presently in use include interim stabilization and covering of excavated material with plastic sheeting in critical areas. Depending upon the length of time a stockpile is to remain and upon the season, the excavated material might require a perimeter ditch as well as a settling area. As a general rule, stockpiling should not be located adjacent to watercourses.

At the completion of construction, the geoenvironmental inspector must see that any excess material is removed and placed in a more appropriate predetermined location – this does not include flood plain areas. Alternatively, the site engineer may wish to establish better contour and grades in particular areas to help restore natural site drainage. Chemical analysis and suitability for placement or transfer are essential.

Sediment Control and Erosion

Prior to construction, procedures are established for the minimization of erosion and control of any erosion, which might occur during construction. Erosion is minimized by adherence to requirements of the contract documents regarding slippage, clearing of vegetation, stockpiling, stream crossing and restoration.

Silt traps and ditching devices along the construction route, are designed and constructed to help and control erosion that occurs.

Placement of Soil in Woodlots

Frequently soil material excavated from a right-of-way is placed on either or both sides of it. The access to woodlots along rights-of-way should be restricted by the use of snow fence and educational programs. Soil analysis and approval are required.

Stream Crossing

Stream crossing and the amount of stream bank modifications should be kept to a minimum. Necessary crossings should be at right-angles or as close to right angles as possible. Such features as diversion culverts and fill material must be suitable for the application and properly positioned. Backfill must be satisfactory to prevent erosion and washout. The crossing should be planned to coincide with low flow periods and the work staged to minimize the time of soil exposure.

Discharge to a Watercourse

Any water that is collected or removed from trenches in the construction zone must not be discharged directly into a watercourse until adequate treatment has been provided to remove suspended silt, clay and other contaminants. If treatment prior to discharge into a watercourse is not considered practical, consideration should be given to another means of disposal such as discharging on grassy areas sufficiently remote from the site. Where the water is carrying heavy silt loads it should be directed to sedimentation ponds, constructed for silt control purposes. In all cases approvals for collection, treatment and discharge should be reviewed.

Groundwater

On certain projects, it may be necessary to deal with a shallow perched groundwater level. Where an excavation intersects a shallow perched water table, extra precautions are necessary to ensure that the sewer is impermeable in order to minimize the possible effects of leakage and groundwater contamination, plus groundwater lowering effects. Water pumped out for purposes of construction dewatering should not be discharged directly to watercourses but handled as outlined above. Ensure that there is a minimum disturbance of areas of groundwater seepage, particularly near steep slopes.

Dust Control

Associated with most construction projects is the problem of dust control. Depending upon the time of the year and the weather conditions, conditions may become an annoyance. A general rule of thumb to follow is that if the construction is an urban area, particularly in the summer, periodic spraying with water or the application of calcium chloride is a good preventative measure.

Hydrostatic Testing

Depending on chemical analysis results water used in hydrostatic testing of a facility should not be discharged directly to watercourses. It is recommended that the discharge be buffered with a holding lagoon and/or a discharge pipe and diffuser properly positioned in relation to the receiving waters.

The rate of discharge to and from the holding lagoon to the watercourse must be regulated in order to preclude erosion and sedimentation in the watercourse. Water containing suspended silt, clay and other contaminants should not be discharged to the watercourse without adequate treatment.

Restoration

One of the most important aspects of any construction project is the site restoration.

Prior to construction, procedures should be defined for the interim such as mulching, use of netting, seeding of grasses and long term rehabilitation such as replacement for vegetation) of areas disrupted by the construction. As far as possible, disturbed terrain should be returned to its original contours. However, the drainage in certain areas can perhaps be improved upon where considered desirable.

Conclusions

If protective environmental controls are adopted, then the planning, design and implementation of a project can proceed with controlled and favourable environmental results. The contractor and engineer must be ready at each stage to implement various mitigation and control measures for the site characterization and changing site conditions.

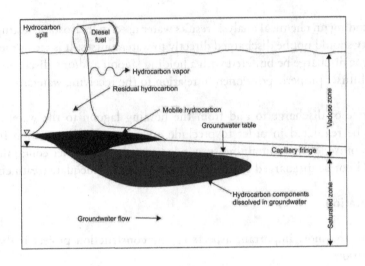

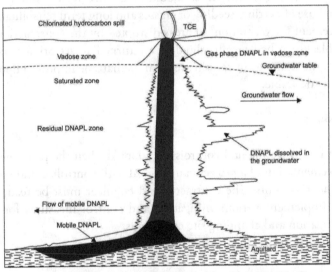

35.1 Hydrocarbon Spill, upper LNAPL and lower DNAPL Components

Chapter 36

GEOENVIRONMENTAL SITE REMEDIATION

The goals of geoenvironmental site investigations and feasibility studies, environmental site remediation, and hazardous waste cleanup projects is to obtain sufficient site information to consider and to select practical remedial alternatives. Contaminants may be separated into five contaminant groups as follows:

- Volatile organic compounds (VOCs)
- Semi volatile organic compounds (SVOCc)
- Fuels
- Inorganics (metals, radioactive elements)
- Explosives

The specific site conditions and contaminant characteristics may limit the applicability and effectiveness of any of the several available treatment technologies listed below.

Soil, Sediment and Sludge

Insitu Biological Treatment
    Biodegradation
    Bioventing

Insitu Physical/ Chemical Treatment
    Soil Flushing
    Soil Vapor Extraction

Insitu Thermal Treatment
    Insitu Vitrification

Exsitu Biological Treatment (Assuming Excavation)
    Composting
    Continuous Solid Phase Biological
    Landfarming
    Slurry Phase Biological Treatment

Exsitu Physical/Chemical Treatment (Assuming Excavation)
    Chemical
    Dehalogenation (BCD)
    Dehalogenation
    Soil Washing
    Soil Vapor Extraction
    Solvent Extraction

Exsitu Thermal Treatment (Assuming Excavation)
    High Temperature Thermal
    Incineration
    Low Temperature Thermal
    Pyrolysis
    Vitrification

Other Treatment
    Excavation and Off-Site Disposal
    Natural Attenuation

Groundwater, Surface Water and Leachate

Insitu Biological Treatment
    Co Metabolic Treatment
    Nitrate Enhancement
    Oxygen Enhance. with Air
    Oxygen Enhance, with H_2O_2

Insitu Physical/Chemical Treatment
    Air Sparging
    Dual Phase Extraction

Hot Water or Steam Flush
Passive Treatment Walls
Slurry Walls
Vacuum Vapor Extraction

Exsitu Biological Treatment (Assuming Pumping)
Bioreactors

Exsitu Physical/Chemical Treatment (Assuming Pumping)
Air Stripping
Liquid Phase Carbon
UV Oxidation

Other Treatment
Natural Attentuation

Air Emissions/Off-Gas
Biofiltration
High Energy Corona
Membrane Separation
Oxidation
Vapor Phase Carbon

Landfills

Earth containment facilities, in various forms have been used for many centuries. Prehistoric people built their village on concentric soil mounds, allowing for ponding of village wastes in the depressions between mounds. Farmers impounded water during periods of rainfall to provide a reservoir for use during drier times. As industrialization occurred, many earthfill containment facilities were constructed to retain raw materials and/or waste products. Most of these containment facilities were not designed and mostly none were lined to prevent leakage of wastes into surrounding sensitive environment. New waste containment facilities must meet stringent government requirements, often involving elaborate double composite liner systems. Many existing facilities must either be remediated or cleaned up, and closed or retrofitted with pollution reduction/prevention systems and monitored to ensure that current legal requirements for non pollution are met.

A number of types of containment facilities are currently used by industry and government to hold both clean and waste materials. These include municipal drinking water reservoirs and sewage lagoons, petroleum storage tank berms, dredged material containment facilities generally can be categorized as either surface impoundment usually containing liquids, or landfills containing solid materials. They provide either temporary storage as in the case of some surface impoundments, or permanent storage i.e. surface impoundments and landfills.

Although impoundments may be used to retain clean drinking water, impoundments and landfills are used principally to contain wastes, including hazardous wastes. According to U.S. Environmental Protection Agency (U.S. EPA) surveys almost 50 percent of hazardous wastes were disposed of in surface impoundments in the 1970s, while 33 percent were placed in landfills.

Landfills designed to contain municipal solid waste (MSW), ash fill, and hazardous waste, constitute a special class of containment facility that is heavily regulated by federal, state and provincial laws. Requirements are specified regarding types and conditions of acceptable waste materials, methods for waste placement and compaction, lining system design and construction, leachate collection systems, and monitoring both during and after active operation. There are requirements for documenting design and construction plans, activities, and emergency operations with regulatory agencies. Various permits must be obtained prior to construction of these facilities. The regulatory and permit requirements are evolving with time.

The basic components of a landfill consist of a lined basin that may be either above or below the existing ground surface, the waste, a gas collection system, and a final cover system. During filling of the facility, daily and intermediate covers are placed on the waste to minimize rainfall infiltration, control nuisance vector populations (rodents and insects), and provide a working platform for future equipment operations.

The types of waste and their characteristics vary significantly, within a single landfill. Such materials as municipal garbage, vegetation, construction debris, tires and old appliances may be found in almost any municipal solid waste landfill. Hazardous waste landfills contain many types of concentrated, toxic, flammable corrosive, or otherwise hazardous materials. Other materials that are not hazardous but have been contaminated by spills or other exposure are also placed in hazardous waste landfills, such as soils contaminated with PCBs and building materials contaminated with asbestos. Materials placed in these landfills are normally required to be in solid form, thus requiring hazardous liquids to be solidified by special procedures before they are accepted for landfilling. Because the contents of landfills vary, the leachate generated by percolation of water through landfills can have very different chemical compositions and may require different lining systems and handling techniques

Regulations for landfill lining systems can vary widely from country to country, from state to state and province to province. Most regulations now require that all landfills be lined. This has a major impact on numerous existing landfills, causing closure of many small municipal landfills. The current minimum containment requirements imposed by the United States Environmental Protection Agency (U.S. EPA) require a double liner system with leachate collection and leak detection systems for hazardous waste and a single composite liner with a leachate collection system for non hazardous waste.

The more stringent liner, cover, and monitoring requirements, coupled with strong public concerns to new waste facilities and escalating costs, are leading to renewed consideration of recycling, waste minimization, incineration, etc. in the future.

Because of land availability and costs, government regulations, and public opposition, it has become difficult to obtain permits for new landfills (green field landfills are currently undeveloped lands). This has resulted in the expansion of existing landfills and/or costly shipping of waste to distant landfills. Landfills expansions may take either two forms: vertical expansion (building on top of an old landfill) or lateral expansion (building beside an existing landfill). Another result of the current regulatory and public atmosphere is that many of the entities involved in waste disposal often exceed minimum lining requirements in an effort to appease public opposition and to remain in compliance with anticipated stringent regulatory requirements. Geoengineering input is essential at all stages of landfill planning, design, construction, operations and closure.

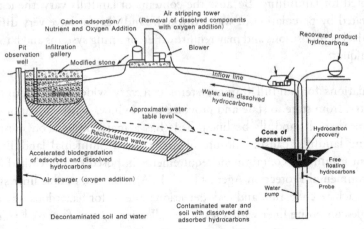

36.1 Hydrocarbon Spill and Site Remediation

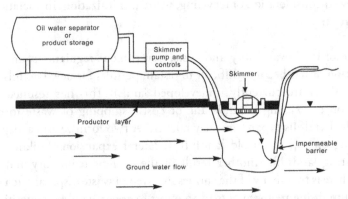

36.2 Hydrocarbon Spill Recovery

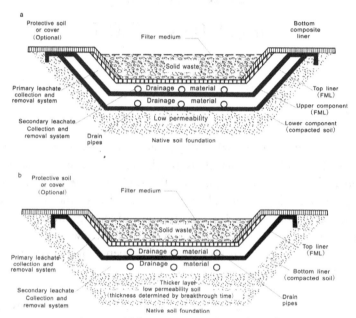

36.3 Landfill Design and Construction

Chapter 37

GEOSCIENCES IN MEDICINE

In North America our geoengineering approach is to focus on environmental and health considerations which deal with localized, regional or site specific soil and groundwater contamination. The typical scope of a modern site investigation includes micro scale consideration of both the geoengineering properties of the subsurface elements, and how to manage any contaminated soil and groundwater in a manner consistent with laws and regulations, protective of human health and the environment, and in the most cost effective manner possible.

In the macro scale approaches to geosciences medical researchers have found that the geomagnetic fields of the earth can affect our health. When humans are separated from naturally occurring electromagnetic fields (EMFs) from the Earth it can have a negative impact on health because they function as a kind of energy nutrient. It is observed by medical researchers that the amount of time people now spend in buildings and vehicles, which are tightly closed spaces, reduces their exposure to the geomagnetic field of the Earth and it may adversely interfere with their health. Doctors call this condition magnetic field deficiency syndrome stating that it can cause headaches, dizziness, muscle stiffness, chest pain, insomnia, constipation and general fatigue. Furthermore the beneficial geomagnetic waves from the Earth can be easily deflected by certain human made technological devices and overpowered by human made stress producing electromagnetic fields such as associated with surrounding transmission lines, electrical wiring, electrical and electronic equipment, microwaves, motors, and the like. Electromagnetic changes in the environment can adversely affect the energy balance of the human organism and health.

As a general rule stronger EMFs generated by human made technological devices or installations tend to be harmful to human health versus naturally occurring weaker EMFs.

Geopathic Stress

Some energies from the Earth can have a carcinogenic effect on humans. Magnetic radiations from the Earth connected with geological fractures and subsurface aquifers when situated below buildings can have deleterious effects on the occupants.

In 1929 the concept of geopathic, or pathogenic, influences from the Earth, was studied in Germany when Baron von Pohl surveyed the community of Vilsbiburg. It had 565 houses, 3,300 residents, and an unusually high rate of cancer. Von Pohl acted on speculation inspired from other surveys that showed a strong correlation between major geological faults in the city and those districts that had the highest cancer mortality rates. His early conclusion was that an unknown but noxious radiation emanating from the Earth faults might be an important and overlooked contributory cause of the cancers.

Von Pohl located all the major subsurface aquifers which are lying at a depth of 44 to 50 metres with a width of 3 to 4 metres under Vilsbiburg, then mapped their courses onto the city street plan. This was cross referenced with the residences of the 54 cancer fatalities to find a startling conclusion: *"The completed check of my map confirmed all the beds of the 54 cancer deaths were where I had drawn the radiation currents,"* von Pohl wrote in 1932 in his classic Earth Currents: Causative Factor of Cancer and other Diseases.

Eighteen months later, von Pohl returned to Vilsbiburg and found that the beds of another 10 cancer mortalities were situated directly over crossing underground streams. The beds in these cancer houses were situated over what von Pohl characterized as geopathic zones marked by dangerous radiation lines. Geopathology is essentially unknown in North America but since von Pohl's work it is the subject of more research, empirical investigation, and medical recognition in his native Germany.

Von Pohl presented more cases in which rapid, perhaps miraculous, cures of numerous complaints, from insomnia to heart spasm, were achieved simply by moving the bed out of the geopathic zone situated below the house. In the town of Stettin, von Pohl's colleague, Dr. Hager, collated the details of 5348 cancer deaths over a 21 year span and found in each case that a subsurface aquifer was located under the cancer patient's house. *"Medical science has now a preventative measure which did not exist previously,"* noted von Pohl. *"If one makes sure one's bed does not stand above a strong underground current and one tries not to work above these underground currents, one should significantly reduce cancer."*

Geophathic stress is mostly unacknowledged in North America as a contributing carcinogenic influence. However, European physicians are more aware of this factor as a contributing cause for cancer. According to German medical researcher physicians 93% of all patients with a malignancy have been exposed to geopathogenic influences. As a standard aspect of cancer treatment, it is advised to remove the patient from sites of geopathogenic exposure, noting that such sites may be identified by way of dowsing or a magnetometer which is capable of registering abnormal magnetic fields.

There are at least two theories used to explain the origin of geopathic stress. One theory says it comes from underground; another says from above, through cosmic rays. According to a biophysics researcher and inventor in Massachusetts, the cause of geopathic stress is localized magnetic anomalies - unusual, sudden and local changes that can upset delicate human physiological balance and create problems.

To prove one hypothesis, the researcher invented a device called a geomagnetometer, which takes precise, local, magnetic field readings in a suspected geopathic zone such as a bedroom. The device then prints a 3-dimensional picture of the disturbed magnetic field. It was found that geopathic stress consists of several factors, maybe as many as 25, but the main factor is a disturbed magnetic field. Here the natural homogeneous magnetic field meets with or turns into a non homogeneous field, resulting in a disturbed zone. These geomagnetic anomalies act upon the human organism as stimuli of a localized and chronic nature and, depending on the intensity and length of exposure, lead to impairment of health.

The focus in the medical research model is a geological factor. It might be a fissure, or fault; the hydrogeologic activity of an aquifer is a secondary factor. The geological factor produces a disturbance in the local magnetic field, which registers as a sharp, sudden, vertical fluctuation in an otherwise smooth, steady field. It is the degree of change in the magnetic field, and the part of the human body (lying in bed) affected by this change, that is of importance. Researchers state that different parts of the same body have different intensity lines and this disturbed zone, where the highest gradient is shown, is often the site of the cancer or illness.

A second theory explaining the possible origin of geophathic stress points to the sky specifically to a vertical field going from the ground up to the sky, suggested by researchers at the Institute of Bioenergetic Medicine in Dorset, England. The vertical field may comprise rays from deep space. It has been assumed that the vertical field holds the same intensity throughout fairly large areas, but actual readings with the magnetometer indicate wide variability in field strength within short distances. This variability is the origin of geopathic stress. The body is unable to adapt to large changes in the field strength within short distances (as few as 150 mm), so its balance or homeostasis, is upset, immune functions can be depressed, and chronic illness can result.

In 1971, the theory of geopathic stress was supported by research showing that water flowing underground, especially subterranean aquifers that cross, produces measurable increases in magnetic anomalies. These subsurface conditions are reported to increase electrical conductivity in the air and soil, and other physical changes. While the changes may be small, though measurable they are still capable of contributing to the development of illness, including cancer. One large scale study by the U.S. government reported that geopathic stress may be a factor in between 40 and 50% of all human cancers and account for between 60 and 90% of all cancers attributed to environmental radiation.

Feng Shui

Some medical practitioners state that geopathic stress in home and work environments is a key factor in some patients to respond to treatment, including alternative medicine. Geopathic stress zones may be small, but shifting a desk or bed a few metres in geopathically troubled rooms, can make the difference between cancer and no cancer for the susceptible individual. The growing popularity of feng shui, the Chinese science of landscape interpretation and building exterior and interior design, is bringing the concepts of geopathic stress to a larger audience of building planners and designers in the Western countries. It may become a common and necessary survey in geological mapping and geotechnical site investigations.

Chapter 38
COMPUTER ANALYSIS AND MODELLING

Modelling forms an implicit part of all geoengineering design but many engineers are not aware of the fact that they are making assumptions as part of the modelling or of the nature and consequences of those assumptions. Many engineers make use of numerical modelling using sophisticated software programs but may not have considered the approximations and assumptions that are implicit in that modelling and the nature of the constitutive models that may have been used. These are new possibilities and implications with physical modelling either at single gravity or on a centrifuge at multiple gravities.

Geoengineering is fundamentally concerned with modelling. Engineering is concerned with finding solutions to real problems. We need to be able to understand the problem and identify the key features which need to be modeled. These features need to be accounted for and included in the design. One aspect of engineering judgement is the identification of those features which we believe it is safe to ignore.

A model is an appropriate simplification of reality. The skill in modelling is to identify the appropriate level of simplification—to recognize those features which are important and those which are unimportant.

Empirical Models

Although the preference is for models which have a sound analytical or theoretical basis there is a long history of empirical modelling in geotechnical engineering. The dictionary tells us that empiricism rests solely on experience and for known soils which the empirical rules were originally generated, such procedures may be representative. However, there is less accurate representation attached to extrapolation to new geological environments and new soil types, where the application of a more complete underlying theoretical model appears to have greater prospects of success.

Theoretical Models

There are two reasons for the continued successful application of empirical models. On the one hand, geotechnical design cannot be halted while more rigorous models are developed. Experience provides a reassuring mode of proceeding. On the other hand, even when accepted theoretical models exist it may not be easy to apply them for the actual boundary conditions of a particular problem. Theoretical models can be seen as solutions looking for problems to which they can be applied. An initial step is often to assess how the observed soil behaviour can best be fitted into the framework that the theoretical model imposes. Once a theoretical model has been formulated there are two possibilities for its application; either the boundary conditions of the problem can be arranged in such a way that an exact analytical result can be obtained; or a numerical solution is required.

Numerical Modelling

Understanding the controlling physical constraints on each problem is important. Within an understanding of the physics there is usually a need to idealise the material characterization and the representation of the boundary conditions of the problem in order that a solution may be obtained. Exact, closed form solutions are in general only obtainable for a rather limited set of conditions. There is a temptation to convince oneself that a problem can be fitted into one of these limited sets because of the ease with which a solution may thus be obtained. It is always necessary to consider whether the arranging of the problem to fit these constraints removes any key characteristics of the problem being considered. Where the departure from the ideal situation is too great there is the possibility of using numerical techniques to obtain a solution, thus an underlying simple and widely accepted theoretical description of the physics of the problem on a local scale but using the numerical approximation to allow realistic boundary conditions to be accommodated.

Numerical solution usually implies the replacement of a continuous description of a problem by one in which the solution is only obtained at a finite number of points in space and time. Certain behaviours of soils in earthquakes, such as the development of quicksand or sand liquefaction, contribute much to the damage to buildings. Studies of soil dynamics in relation to earthquake damage have been undertaken by computer modelling. Influences of machine vibrations are also studied.

Pile driving also involves soil dynamics and wave equation approaches have been developed based on computer modelling of soil reactions during pile driving.

Computer modelling of complex soil mechanics problems by finite element analysis is widely used. This requires mathematical modelling of soil strength and volume change behaviour, which is difficult as these tend to be discontinuous functions. Emphasis has been directed toward refined laboratory testing to define idealized constitutive equations to describe soil behaviour under widely varying stress environments. Similar finite element models are used for groundwater flow and contaminant migration

Constitutive Modelling

Numerical modelling is not used in order to manage irregular or non ideal boundary conditions. More serious are the idealizations of material behaviour that are necessary in order that simple theoretical models can be developed. Elasticity is convenient because of the wide range of analytical results to which access can be gained for elastic materials.

The nonlinearity that is observed in soil behaviour is usually an indication of plasticity. These are permanent, irrecoverable changes in the fabric of soil. The simple effects of soil plasticity on the response of a geotechnical structure is provided by the pattern of deformation beneath a footing on a linear elastic soil and on a rigid perfectly plastic soil. The elastic material stays together and a movement in one location is felt at great distance. The footing produces gradients of deformation, and strains, to great depth. The plastic material is considered to come apart into separate blocks of soil as it gradually forms a failure mechanism. The displacements are contained within this failure mechanism. The displacements only occur at the boundaries between the sliding blocks, and they are infinite. Elements of soil at depths below the mechanism are unaware of the footing, that is, the soil has been assumed to be rigid before failure.

Physical Models

Physical modelling plays an important role in the development of geotechnical understanding. An experiment is a physical model intended, if it is a good model, to advance our confidence in some supporting theoretical model which the experiment was designed to investigate. The physical model forms the observation of a reflective practice whereas the theoretical model forms part of the prediction.

Physical modelling is performed in order to validate theoretical or empirical hypotheses. Geotechnical construction is also physical modelling. Geotechnical design makes hypotheses about expected behaviour which may be tested in greater or lesser detail depending on the extent to which the response of a geotechnical system is observed. At the very least there will be a binary observation; has the geotechnical structure failed? A failure will be a pretty clear indication of inadequacy in the supporting models. If the designer is less confident in the supporting design models then more extensive observation - for example of displacements or pore pressures - may provide better information about the way in which the geotechnical materials are in fact behaving. Reflection on these observations provides the insight for improved future design or modelling.

Laboratory testing of small elements of soil (for example in triaxial apparatus, shear box, etc.) and insitu field testing (geophysical testing, penetration testing, pressuremeter testing, load tests, etc.) presuppose some model for the way in which the subsurface is going to respond.

Geological Model

For geology models, the emphasis is on the mechanical aspect of geotechnical modelling. The questions to be answered relate to the engineering characteristics and properties of the geotechnical materials. These materials have been placed either by geological and geomorphological history or by man. Knowledge of this history can help to focus our ideas about the likely mechanical characteristics and a geological or stratigraphic model is usually recommended as a precursor and main feature of the geotechnical model.

A reasonably well developed geological model can lead to economy and efficiency in subsequent site investigations to determine quantitative properties of the ground. Parallels can be drawn with past experience and with adjacent sites with similar geology. The expected properties, the nature and mineralogy of soil particles, the appropriate constitutive models which may in some way predefine the insitu or laboratory testing, and the likely pitfalls can be predicted. For example, fractured rock associated with faulting or irregular buried erosion features in weaker rocks may be anticipated. Although geophysical techniques can be used to obtain an overview of the structure of the ground, detailed knowledge usually comes from discrete boreholes. A geological model is necessary to be able to propose continuity or lack of continuity of stratigraphy between boreholes. The ground is usually not homogeneous. Vertically the inhomogeneities may primarily result from depositional layering; different rock layers at one scale with spacings in the order of metres; varves resulting from seasonal variations in sediment transport and water velocities at another with spacings in the order of millimetres. Horizontally there may also be variations. The geological model can help to understand the reason for and the nature of the spatial variations.

At the simplest level, the boundary between the soil-like materials which are expected to deform and control the behaviour of the geotechnical system, and the rock like materials which are anticipated to be more or less rigid and possibly impermeable in comparison, is important in defining the extent of the ground that needs to be modeled either physically or numerically.

Classification Model

Before any tests to determine the mechanical properties of the soils are performed a classification model is used for which the soils that are encountered are placed into categories. Samples have been recovered; they are possibly disturbed but one can from simple visual inspection categorize the soil as broadly gravelly or sandy or silty or clayey. Particle size distributions can be obtained which confirm the visual classification. These distributions are a simplified model of the soil in which the soil particles are replaced by equivalent spheres. For a soil with particles large enough to sieve, the size of the equivalent spheres is defined by the size of the mesh spacing through which the particles—of whatever actual shape—will fit. For a soil with finer particles, Stokes' law, which describes the terminal velocity of spheres falling through a viscous fluid, is used to define the size of the spheres to which the actual soil particles in their rate of descent through the fluid are equivalent. In addition, some assessment may be made of the typical particle shapes, particle mineralogy, void ratio, water content, liquid and plastic limits, plasticity index.

These classification and index property tests are simple standard tests which can be rapidly performed with low cost equipment but whose results categorize the soils. They make up a classification model which is useful for sharing information across different sites and can provide a basis for other estimates of soil properties. If soils from two sites have similar index properties and similar particle characteristics and similar geological histories, then it is to be expected that other mechanical properties may fit into a consistent pattern across the sites.

Conclusion

Various modelling types are in regular use by geotechnical engineers, but mostly theoretical, numerical, classification and physical modelling.

The equations of equilibrium and of strain compatibility for a continuous material are well established. In order to analyse the deformations of a geotechnical system it is necessary to provide a link between stresses and strains in the form of a constitutive model. For the mechanical behaviour of soils simple linear elastic or perfectly plastic models are likely inadequate in detail - though they may be appropriate in some circumstances. The alternative possibilities of constitutive model may be more applicable to soils. There are often several ways in which the same experimental observation can be modelled.

The nonlinearity of most constitutive models makes it essential to use numerical procedures to obtain solutions to boundary value problems - the behaviour of complete geotechnical systems of interest to geotechnical engineers. Numerical analysis is a tool to be used and the emphasis is not on the theoretical basis of numerical analysis. The power of computers available to all geotechnical engineers has increased. Numerical analysis tools should be used as part of the routine of geotechnical design, incorporating the constitutive models and recognizing the inadequacy of some of the simplifying assumptions that have been used to expedite calculations. With numerical tools there is the need to ensure the reliability of the results for design and construction purposes.

Scaling laws in the design and interpretation of physical models are important. Soils are nonlinear history dependent materials. The understanding of scaling laws for stress related quantities for such materials is not straightforward and the extrapolation of observations made in small scale physical models to the full scale prototype response is simplified if the stress level of the physical models is similar to that of the prototype. This can be achieved by subjecting the physical model to an artificial gravitational field on a geotechnical centrifuge.

Modelling should always be of adequate complexity. Numerical modelling will not be required or appropriate in all circumstances. Elastic analyses are regularly used as part of geotechnical design and for validation of computational tools that are to be used for elaborate calculations. The use of perfect plasticity underpins much of the ultimate limit state design in geotechnical engineering.

One or the main applications of geotechnical modelling, whether theoretical/ numerical or physical, is to assess the consequences of soil structure interaction. Soil structure interaction problems tend to be driven by stiffness or deformation properties of soils. Constitutive modelling of prefailure deformation properties is important.

Similar modelling techniques are used for groundwater movement in hydro-geology and for contaminant migration in geoenvironmental applications.

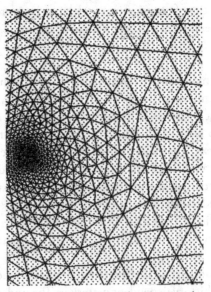

38.1 Subsurface Finite Element Analysis

Chapter 39
COMMUNICATION AND REPORTS

Overview

Along with curiosity and ingenuity every geoengineering problem involves discovery and therefore is a new adventure. Writing is invention, and invention can be gratifying for self and others.

Writing is communication, and an engineer who cannot communicate effectively in verbal and written forms is going to have a difficult time. Effective talking and writing require effective thinking. Writing is a disciplined form of communication.

As they advance in their profession, geotechnical engineers discover that they spend more time using a computer as a word processor than as an instrument for calculations, as calculations now are performed almost instantly by commercially available software while writing takes time. Writing time can be reduced by using stock paragraphs to describe such things as the general geology or standardized test procedures, but the focus of a report will be the specifics of the site findings, conclusions and the report recommendations.

A draft report may be the product of many additions and revisions. Even after it is completed, a report should be thoroughly checked, preferably by a senior associate, to minimize the possibility for misinterpretations, misstatements, errors or omissions.

The relationship between thinking and writing can allow an engineer to arrive at new ideas and concepts during the writing process, which is one reason why advanced degrees, even in engineering, often require a written thesis or dissertation. A thesis or other written report, must follow a rational thought process and procedures from beginning to end.

Geotechnical engineers deal with the world's most abundant construction materials of soil and rock and these materials are the most variable and changeable in response to loading conditions and the environment. Subsurface conditions are non homogeneous and anisotropic. Geotechnical engineers start each project by identifying and characterizing the soil, rock, groundwater and geoenvironmental conditions, and then consider and evaluate the options for design and construction. Engineers become intrigued with this kind of problem solving, partly because it takes them out of the office and into the field. Architects and structural engineers mostly communicate through graphical means. Geotechnical engineers communicate with written reports comprising borehole logs, monitoring wells, plans, sections and laboratory test results.

Engineering and scientific writing follows a narrative style, starting with (1) goals or requirements of the site investigation, (2) a review of literature and other available information, (3) details of the procedures used, (4) results of the investigation, (5) analysis and interpretation of the field and laboratory results and (6) conclusions and recommendations. It may begin to look like boilerplate to the person doing the reading.

Engineers think and write in narrative style, but clients want answers now. A simple and effective style to overcome boiler plate tedium is to include an abstract or executive summary at the beginning of a proposal and report, where it can create interest and quickly convey the most relevant information. The body of the report then can meet requirements of objectivity and the thoroughness of data presentation, thinking and writing.

The first step in the project is to obtain details on the assignment and develop a proposal that is acceptable to the client. As with any project, basic information should be obtained before preparing the proposal, such as:

a) Client Information. This includes the name and address, and telephone number of the client. It may also be appropriate to check the reputation of the client if it is not already known.

b) Project Location. Basic information on the location of the project is required. The location of the project can be compared with known geologic hazards, such as active faults, landslides, or deposits of liquefaction-prone sand.

c) Type of Project. The geotechnical engineer could be involved with all types of civil engineering construction projects, such as residential, commercial, or public works projects. It is important to obtain as much preliminary information about the project as possible. Such information could include the type of structure and use, size of the structure including the number of stories, type of construction and floor systems, preliminary foundation type (if known), and estimated structural loadings.

d) Scope of Work. In order to prepare a proposal that is acceptable to the client, the geotechnical engineer must have specific details on the scope of work. For example, the scope of work could include a preliminary geotechnical investigation to determine the feasibility of the project, compaction testing and observations during grading of the site, and the preparation of foundation design requirements.

e) Conflict of Interest. The geotechnical engineer must investigate possible conflicts of interest. As an example of a conflict of interest, suppose you were hired by a contractor to help prepare a bid for a project. If a different contractor subsequently asked you to also prepare a bid on the same project, you could not accept the assignment because it would be a conflict of interest.

Proposal Writing

Where there is a long term mutual trust and understanding, a geotechnical investigation could initiate with a simple letter or phone call. In other circumstances, the proposal serves two functions, to outline the problem as it was present to the geotechnical engineer and on which the recommendation will be based, and to allow the client to compare costs, timing, and other issues including qualifications and method of approach to a problem. Unfortunately in a competitive market a few clients will be swayed by low cost figures, but the professional engineer is obligated to perform an adequate investigation, and may become liable if the investigation is not keeping with current standards of the profession. The client is not only buying a service, he or she is depending on the experience, expertise, and competence of the engineer to do a proper job.

A written proposal includes a brief job description and location, building loads and other requirements provided by the civil structural engineer and/ or architect, and scope of work. The geotechnical proposal may include a tentative borehole plan, laboratory testing program, and cost estimate, and should be carefully written because if accepted it will be part of a legal binding contract. Any unusual conditions that may be anticipated based on the engineer's experience should be noted, not as a scare tactic, but as a professional reminder of the importance of having a competent geotechnical investigation.

Site Plan Showing Borehole Locations

The scope and intent of a project form the site investigation. Is it for a building, dam, pavement, landfill or contaminant spill. Included should be anticipated final grades and floor elevations, and a preliminary estimate of dead and live loads prepared by a structural engineer. A topographic site survey normally is conducted independently and the information is provided to the geotechnical engineer.

A considerable part of the cost of a site investigation is for boring, and geotechnical consulting firms may calculate a total cost simply on the basis of the number and depths of boreholes. A boring plan is required that will include locations, depths, and tests that are anticipated to be performed in the field and in the laboratory. The proposed testholes may include boreholes, probes and test pits.

Borehole Depths

Anticipated boring depths will depend on the site geology, loads and footprint of the proposed structure, and relationships to topography and other structures. On large projects at least one boring should be deep enough to encounter competent material or bedrock. Other boring depths must be sufficient to extend below the bottom of a bearing capacity failure, which generally is about 1.5 times the width of the bearing area.

A second consideration in addition to bearing capacity is settlement. Because of the decrease in added stress with depth below a foundation, one guideline that is based on elastic theory is to extend borings to a depth such that the added pressure is less than 10 percent of the bearing pressure. The broader the loaded area, the deeper should be the borings.

Some criteria for selecting boring depths and spacings: (a) bearing capacity can depend on the soil, footing width, or depth to a firm layer; (b) settlement depends on pressure, soil type, and the pressure bulb; (c) leakage can readily find paths under an earth dam; (d) sinkhole, debris-filled ravines, and collapsed mine workings are difficult to locate and pose dangers; (e) confined and unconfined aquifers have regional concerns; if containment spills and old landfills have migration concerns.

Borehole Spacing

Borings normally are located at all exterior corners of the buildings to ensure that the same soils exist across a building site, and they may be spaced at no more than 15-30 m (50 – 100 ft) intervals on a grid pattern. A center boring often is extended to greater depth to ensure that there are no radical changes that could adversely affect a construction project. Not only are soils and rocks identified; equally important is the location of the groundwater table, and geoenvironmental constituents.

The boring pattern should be adjusted to reflect geological features observed on the ground or from airphotos, as otherwise important features can be missed by an arbitrary grid pattern.

Insitu tests often are employed, and applications of geophysical methods should not be overlooked when needed between borings.

Contract Performance

After an assignment is awarded or a contract is negotiated the site investigation should be performed in a timely manner, weather permitting. If unusual or adverse site conditions are encountered, that information should immediately be relayed to the client, the structural engineer, and/or the architect, as it could mean a change of plans or of location. The drilling and other field investigations are conducted, samples are taken to the laboratory for identification and analysis, and a geoengineering report is prepared.

Organization of the Report

There are no set rules for the organization of the report and most consulting firms will have developed their own style and preferences. A logical order is as follows:

Title and Signatory Pages

The title page should include the title and date and identify the firm conducting the investigation. The signatory page should include the name and title of the person preparing the report, and the signature and seal of the engineer who will bear the ultimate responsibility for the report.

Executive Summary

An executive summary may include (1) general project description and location, (2) a few words to explain the geology and primary geotechnical concerns, which can include a description of special problems such as expansive clay or random fill, and (3) general recommendations that often will include alternative approaches that can be considered in relation to cost and feasibility.

Body of the Report

- Location and general description of the project. This should include building loads and floor levels that form a basis for the report.
- Site geology with soil and rock. Even though this sometimes is referred to as boilerplate because it can be standardized from files, it is an essential part that shows that the engineer knows the territory. It may be prepared with the assistance of a geologist, and should refer to relevant publications such as available government geological reports and soil surveys. Surface water and groundwater characteristics may be relevant.
- A map showing boring locations.
- Boring logs that include soil and rock identifications and groundwater table, and usually will include Standard Penetration Test and unconfined compressive strength data, moisture contents, liquid and plastic limit data, geological identification, soil colour and gradation, and engineering classification.
- Stratigraphic cross-sections showing boring logs at their proper elevations, geological identifications, groundwater table. Strata are joined by straight lines that are dashed to convey that they are interpolations only.
- Additional test data and graphs, plotted versus depth where appropriate. The data will include results from both insitu field tests and laboratory tests.

- Special considerations: a discussion of groundwater, quick conditions, slope and landslides, expansive clay, collapsible soils, liquefaction potential, sinkhole mines, hazardous wastes, and other potential problems, plus geoenvironmental considerations.
- Summary.
- Conclusions and recommendations.
- An indication of availability for questions, discussions, and additional supplementary investigations.
- Disclaimer: A legal clause indicating that the interpretations and conclusions in the report are based on the available data and boring information, and cannot guarantee conditions that may exist between the borings. Usually prepared by an attorney in conjunction with insurers.
- Every report must be checked. A final step before a report is submitted is to have it carefully checked word for word by an associate or principal engineer. Clarity and accuracy are requirements, and correct spelling is non negotiable. Many offices require that two engineers sign every report.
- Client presentation visit: an optional meeting to review the report and any follow up.
- Invoices for services. This is mailed separately out of consideration for the client's sensibilities.

Daily Field Reports

Daily field reports are common on construction projects. For example, daily field reports are used during earthworks. The report could be prepared by a field technician, engineer, or geologist. The daily field report would typically list the type of equipment and personnel, the area of work the results of field tests, and observations during construction operations.

Daily field reports are recorded on duplicate forms, which are commonly titled, "Report of Field Observations and Testing." The form usually has the company's name, address, and telephone number as well as spaces for the date, writer, project name and number, and client. The form usually has a main section to document field observations and testing. Some common items listed on this section of the daily field report are the contractors' operations, test results, and sketches or diagrams.

In many cases, the daily field report will be an inhouse document that is used to prepare a construction progress report or final project completion report.

Frequently a copy of the daily field report will be given to the contractor, supervisor or client.

The degree of care which goes into a daily field report should not vary depending upon who the reader will be as eventually it may be read in a courtroom.

File Management

Examples of the contents of files include copies of reports sent to the client, daily field reports, correspondence, calculations, notes and maps.

Some engineers believe that all documents, no matter how trivial or redundant, should be saved. Problems could occur if confidential, personal, or unrelated material is discovered in the file. At the other extreme are files devoid of all documents, except for reports issued to the client. This could cause problems because important photographs or testing of data may have been discarded. Neither extreme is desirable.

A file should be divided into separate compartments, such as confidential material, correspondence, testing, photographs, analysis, reports, and maps. Each individual compartment can be scrutinized for inappropriate material. Items that should be discarded include prior drafts of reports and letters, personal documents, rough notes or calculations (but keep the finished draft), and inappropriate or unrelated material. When a project is complete, the project manager should review the file and prepare it for the archives.

In many cases, the daily field report will be an inhouse document that is used to prepare a construction progress report or final project completion report. Frequently a copy of the daily field report will be given to the contractor, supervisor or client.

The degree of care which goes into a daily field report should not vary depending upon who the reader will be as eventually it maybe read in a courtroom.

The writer of a daily field report should prepare the document with the same care as a client's final report. Several rough drafts may be needed before the completion of the final report. A daily field report should not be a hasty summary of events, but a well thought out document that reflects field observations accurately.

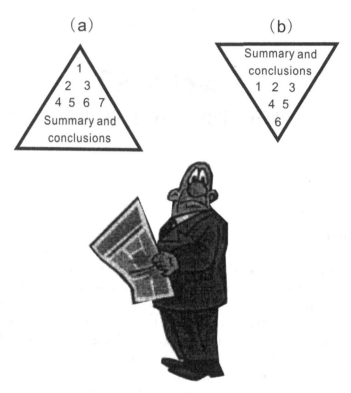

(a) Engineers think and write in narrative style, but
(b) Clients want conclusions and recommendations now.

39.1 Engineering Reports

Chapter 40
INTERNATIONAL SYSTEM OF UNITS

Background

The Système International d'Unités, or as it is known in English as the International System of Units, is often abbreviated to SI. In 1960 it was introduced as a modern standardized version of the various different metric systems. It is the worlds most widely used system of measurement and is used both in the scientific community as well as commerce. SI is used in most countries around the world with very few exceptions, most notably the U.S. which continues to use customary units along with SI. After World War II a number of different systems of measurement were still being used throughout the world. Some of these systems were variations of the metric system, while others were localized customary systems.

The SI system is always evolving. The units stay the same but the definitions may change periodically based on more accurate ways of conveying the same dimension. Every four years in Paris, France the General Conference on Weights and Measures (*Conférence Générale des Poids et Mesures*, CGPM) is held. The goal of the CGPM is to discuss and examine the arrangements required to ensure the propagation and improvement of the International System of Units. Any changes to unit definitions are decided upon at these conferences.

SI measurement is considered to be simple to use because of three main reasons: SI measurement is based on seven different standards called base units, prefixes are used with base units to denote multiples or fractions of the base unit, no fractions are used in SI, only decimals and because prefixes are used there are never long rows of zeros.

Development

The SI is built on seven fundamental standard units often referred to as base units which can be used together to convey any physical quantity. These seven units are the metre, kilogram, second, ampere, Kelvin, mole, and candela which correspond to the respective quantities of length, mass, time, electric current, thermodynamic temperature, amount of substance, and luminous intensity.

The original definition of the metre was based on an international prototype developed out of platinum iridium in the late 1800's. This was updated in 1960 to the current standard which is: *The metre is the length of the path travelled by light in vacuum during a time interval of 1/299 792 458 of a second.* It should be noted that the speed of light in a vacuum is 299 793 458 m/s.

The kilogram unit is based on a platinum iridium artifact developed in the late 1800's. To end the ambiguity associated with the term weight the following declaration was made: *The kilogram is the unit of mass; it is equal to the mass of the international prototype of the kilogram.*

For many years the second was originally defined as a fraction of the mean solar day, but throughout the years it was found that irregularities in the earth's rotation made this definition unsatisfactory. Through recent experimentation it was shown that an atomic standard of time based on the two energy levels of a molecule could be reproduced more accurately. This lead to the current definition of the second, which is: *The second is the duration of 9 192 631 770 periods of the radiation corresponding to the transition between the two hyperfine levels of the ground state of the caesium 133 atom.*

In the late 1940's it was decided that the Ampere would be adopted as the standard unit to measure electrical current and thus the current definition of the ampere was developed: The ampere is that constant current which, if maintained in two straight parallel conductors of infinite length, of negligible circular cross-section, and placed 1 m apart in vacuum, would produce between these conductors a force equal to 2 x 10^{-7} Newton per metre of length.

In SI the standard unit for the measurement of thermodynamic temperature is Kelvin. To which the definition is: The Kelvin, unit of thermodynamic temperature, is the fraction 1/273.16 of the thermodynamic temperature of the triple point of water. It should be noted that since the magnitude of Celsius temperature is equal to that of Kelvin, it is often used in most non scientific usage.

The quantity used by chemists to specify the amount of chemical elements or compounds are now referred to as the amount of substance is officially defined as:

1. The mole is the amount of substance of a system which contains as many elementary entities as there are atoms in 0.012 kilogram of carbon 12.
2. When the mole is used, the elementary entities must be specified and may be atoms, molecules, ions, electrons, other particles, or specified groups of such particles.

Candela, the unit used to measure luminous intensity, derived its name from people using flame or incandescent filament standards to measure luminous intensity. Because of new advances in radiometry a new adopted definition of the candela was determined to be: *The candela is the luminous intensity, in a given direction, of a source that emits monochromatic radiation of frequency 540 x 1012 hertz and that has a radiant intensity in that direction of 1/683 watt per steradian.*

A knowledge and use of both SI units and Imperial units may be required in some countries.

SI Base Units

Name	Symbol	Quantity
metre	m	length (l)
kilogram	kg	mass (m)
second	s	time (t)
ampere	A	current (I)
kelvin	K	temperature (T)
mole	mol	amount of substance (n) number of particles
candela	cd	luminous intensity

Common Standard Prefixes for the SI Units of Measure

Name	Symbol	Factor
nano-	n	10^{-9}
micro-	μ	10^{-6}
milli-	m	10^{-3}
centi-	c	10^{-2}
deci-	d	10^{-1}
deca-	da	10^{1}
hecto-	h	10^{2}
kilo-	k	10^{3}
mega-	M	10^{6}
giga-	G	10^{9}
tera-	T	10^{12}

Conversion Factors from Customary Units to SI for Common Quantities

From:	To:	Multiply by:
Length		
inch (in)	metre (m)	2.54×10^{-2}
feet (ft)	metre (m)	0.3048
mile (mi)	metre (m)	1.609×10^{3}
Area		
square inch (sq.in.)	square metre (m^2)	6.452×10^{-4}
square foot (sq.ft.)	square metre (m^2)	9.291×10^{-2}
hectare (ha)	square metre (m^2)	1.0×10^{4}
acre	square metre (m^2)	4.046×10^{3}
Volume		
liter (L)	cubic metre (m^3)	1.0×10^{-3}
gallon (U.S.)(gal)	cubic metre (m^3)	3.785×10^{-3}
	liter (L)	3.785
cubic foot (cu.ft.)	cubic metre (m^3)	2.832×10^{-2}
Acceleration		
inch per second squared (in/s^2)	metre per second squared (m/s^2)	2.54×10^{-2}
foot per second squared (ft/s^2)	metre per second squared (m/s^2)	0.328
Force		
pound - force (lbf)	newton (N)	4.448
kip (1 kip = 1000 lbf)	newton (N)	4.448×10^{3}

Density		
pound per cubic inch (lb/in³)	kilogram per cubic metre (kg/m³)	2.767×10^4
pound per cubic foot (lb/ft³)	kilogram per cubic metre (kg/m³)	16.01
Flow		
gallon per minute (gpm)	cubic metre per second (m³/s)	6.309×10^{-5}
	litre per second (L/s)	6.309×10^{-2}
gallon per day (gal/d)	cubic metre per second (m³/s)	4.381×10^{-8}
	litre per second (L/s)	4.381×10^{-5}
Velocity		
kilometre per hour (km/h)	metre per second (m/s)	2.778×10^{-1}
foot per second (ft/s)	metre per second (m/s)	3.048×10^{-1}
mile per hour (mi/h)	metre per second (m/s)	4.470×10^{-1}
	kilometre per hour (km/h)	1.609
revolution per minute (rpm)	radian per second (rad/s)	1.047×10^{-1}
Pressure		
pascal (Pa)	newton metre squared (N/m²)	1
atmosphere (atm)	pascal (Pa)	1.013×10^5
bar (bar)	pascal (Pa)	1.0×10^5

Multiplying factor			SI Prefix	SI Symbol
1 000 000 000 000	=	10^{12}	tera	T
1 000 000 000	=	10^{9}	giga	G
1 000 000	=	10^{6}	mega	M
1 000	=	10^{3}	kilo	k
100	=	10^{2}	hecto	h
10	=	10^{1}	deca	da
0.1	=	10^{-1}	deci	d
0.01	=	10^{-2}	centi	c
0.001	=	10^{-3}	milli	m
0.000 001	=	10^{-6}	micro	μ
0.000 000 001	=	10^{-9}	nano	n
0.000 000 000 001	=	10^{-12}	pico	p
0.000 000 000 000 001	=	10^{-15}	femto	f
0.000 000 000 000 000 001	=	10^{-18}	atto	a

40.1 SI Prefixes

Chapter 41
QUALIFIED PROFESSIONALS

The pure and applied sciences used in geoengineering are formal to civil engineering and mining qualifications. The essential education is in physics, chemistry, and mathematics followed by university courses in geology, civil engineering, geophysics, rock mechanics and rock engineering, soil mechanics and foundation engineering, hydrogeology, environmental engineering, geochemistry, construction quality assurance monitoring, materials laboratory testing, instrumentation and monitoring, computer analyses and modelling, and specialty courses. Survey camps and cooperative work placement programs are ideal.

The geoengineering training must include field experience, with juniors and intermediates spending most of their time in the field, the longer the better, and seniors and principals never overlooking the need to observe and survey every site prior to commencing investigations, design and construction. Field experience supports engineering judgement, the most prized possession of all members of the geoengineering profession. Knowledge from experience is not always common, and intuition by nature is very important versus factual knowledge derived from formal education and simply long term observations. The difference between practical experience, and an intuitive and developed appreciation of geoengineering concepts is considered to be indeterminate. Some practitioners may not acquire intuitive judgement after a lifetime of field experience, others acquire it easily and early in their careers.

The geoengineering concepts and practices outlined in these chapters are basic to a full appreciation of the role that the specialist geoengineering team can and must play in the planning, design and construction of civil engineering and mining works. The concepts and practices are relatively straightforward and simple, and their application requires no unusual skills. The ability to look below the ground surface by subsurface investigative methods; to realize that the surface conditions may be vastly different from the subsurface conditions to varying depth layers; to keep in mind the ever present possibility of faults and voids and groundwater movement and contaminant migration; to visualize the significance of unusual topographic and terrain features; and the ability to relate these to geologic processes of the past – all these are characteristics of observation in which the geoengineer can be trained, or self trained, and rapidly advanced by curiosity, intuition and ingenuity.

Ground failures and disasters are rarely caused by a lack of precision or miscalculation, but more often by neglect, ignorance, or over specialization. The ground condition is unforgiving for those who turn their back on it. In geoengineering it is wise to be cautious, because subsurface conditions are non homogeneous and anisotropic, and difficult to predict accurately in advance of construction. Subsurface information at test points should be accurate, but changes in between must be anticipated by landowners. Certain recommendations and decisions are made and cost estimates are based on the best available information which can be obtained within a limited time and, more importantly, a limited budget. Subsurface uncertainties are reduced by all available means to a level considered to acceptable, but even then, construction and mining must provide for the unanticipated anomalies.

The Consulting Engineer

A consulting engineer may be a sole professional practitioner or an organization of many professional and technical people whose prime purpose is to offer clients consultative services.

Consultants are not committed to any manufacturer or contractor and can therefore offer clients an unbiased overview through independent study. The only material compensation for services rendered is a fee.

Consulting engineers are professionally uninhibited, innovative, broadly experienced and objective. They are open to alternatives and are oriented to their clients' needs - functionally, economically, environmentally, and aesthetically - in a timely fashion.

41.1 Consulting Engineer Role and Responsibility

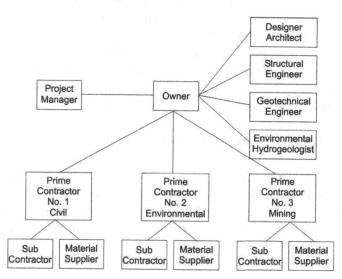

41.2 GeoEngineering Project Management

Chapter 42

REGULATORY FRAMEWORK AND SAFETY

Building Codes

Many North American cities and other organizations such as transportation and highway departments have established tables of safe bearing capacities on the basis of experience with the local soils.

Building codes are a valuable guide, but are oversimplified. For example, they typically specify a value or a narrow range of values for safe bearing capacity for each class of soil without giving any consideration to width, shape, and depth of a foundation, or to the proximity to other structures, and even the relation to a groundwater table. The foundation width is particularly important because wider foundations confine the soil and allow a higher bearing pressure. Building codes therefore may include an option that design can be performed by a registered professional engineer.

The Ultimate and Serviceability Limit States

In the ultimate limit state (ULS) design of foundations, the bearing capacity of the foundation is compared to the actions from the various loading conditions. By introducing partial safety factors both on the actions (by multiplication) and on the bearing resistance (by division), it can be verified that the foundation is sufficiently far away from the failure conditions. In a similar way the foundation can be verified against structural failure.

In the serviceability limit state (SLS) design of foundations one must prove that settlements and deformations of shallow foundations and piles due to the characteristic loads are compatible with the building.

Total Safety Concept

In many countries, until recently, the building foundation bearing capacity has been determined on the basis of a total safety concept. The allowable bearing capacity of a shallow and pile foundation is determined by dividing the ultimate bearing capacity by a total factor of safety (F.S.) for example, F.S. = 2.0 is used in many countries.

Excavation and Trench Base Safety

One of the most hazardous areas on any job site is in a tunnel, a shaft, a trench, an open excavation near a vertical wall, or the crest of a steep slope. Many deaths and injuries have been caused when soil and rock have caved in, covering people working in the excavated areas.

As previously mentioned in North America federal governments and most provinces, states and territories have enacted laws designed to insure that contractors take proper precautions to protect workers in these situations. Occupational Health and Safety (OHS) organizations have issued regulations that specify precisely how most aspects of construction activities are to be conducted with human safety in mind. The excavation safety regulations cover shoring, slope angles, stabilization and protection of slopes, and protection systems for the surface areas around excavations.

A copy of OHS regulations may be found in most large libraries. The various construction trade associations usually have copies of OHS regulations pertaining to their particular trade available for their members' use. Additionally, most trade associations regularly conduct safety classes for both workers and management personnel, which address the content of the OHS regulations. The requirements are specific to a jurisdiction.

Designers and contractors should consult the entire OHS construction safety regulation text or a safety professional for additional construction safety information.

Geoengineering is most concerned with excavation, trenching, and shoring.

The extent of OHS regulations would fill volumes, so only general guideline regulations most pertinent to excavation safety will be outlined here.

1. Generally, excavations more than 1.2 m deep must be sloped, shored, or have worker-protective shields installed. If the sides are to be sloped, the required angle of slope depends on the type of soil encountered during excavation. The required angle of slope varies for soil types.

2. Means of egress from excavations must be provided. Ladders or other means that would allow workers to exit the excavation must be provided at least every 8 m in trenches more than 1.2 m deep.

3. Excavations over the specified maximum depths must be approved by a registered professional engineer.

4. Material may not be stored near the edge of any excavation. Overloading of the surface area near the excavation must be avoided.

5. The rim of the excavation must be protected by suitable barriers designed to keep workers from falling into the excavation.

6. The excavation must be protected from water accumulation or erosion and equipment vibration. The loads imposed by water or equipment vibration may cause either sloped or shored excavations to become unstable. If the strength design of the slope or the shoring is exceeded, a potentially disastrous caving or shifting of the slope may result.

TYPICAL ALLOWABLE SLOPES	
Soil or Rock Type	Maximum Allowable Slopes (H:V) [1] For Excavations Less Than 6 m Deep
Stable Rock	Vertical (90°)
Type A [2]	¾:1 (53°)
Type B	1:1 (45°)
Type C	1 ½:1 (34°)

Notes:

1. Numbers shown in parentheses next to maximum allowable slopes are angles expressed in degrees from the horizontal. Angles have been rounded off.

2. A short term maximum allowable slope of 1/2 H:1V is allowed in excavations in Type A soil that are 3.7 m (12 ft) or less in depth. Short term maximum allowable slopes for excavations greater than 3.7 m (12 ft) in depth shall be 3/4H:1V (53°).

3. Sloping or benching for excavations greater than 6 m (20 ft) deep shall be designed by a registered professional engineer.

As previously stated, the safety items presented above are only a part of an entire construction safety program. The proper OHS regulation for a specific jurisdiction should be consulted for more detailed information for designers and constructors.

It is important to understand that although it took powerful heavy equipment to open an excavation, the walls of the excavation may, in reality, be quite fragile, therefore adequate safety precautions must be taken prior to entry by workers and equipment.

Worker Safety in Trenches

Even partial burial of a person as a result of a trench cave in can be fatal because the impact from caving soil expels air from the lungs, and in order to inhale, the victim must overcome passive resistance from soil that is composed of clumps and has a very high friction angle. Oxygen will be needed unless the person can be quickly freed. Prevention of blood circulation in the legs can lead to future medical problems. Rescuers who enter the trench run the risk of being caught in another cave in, compounding the tragedy. The only safe procedure is prevention, and safety regulations now require that workers in trenches with vertical walls be adequately protected by shoring or by a steel trench box that is pulled along as pipe sections are installed.

Protection of Facilities

It is desirable and usually necessary to protect the completed primary underground structures associated with power plants, pumping stations, dams, tunnels, shafts, waterways and slopes, so that they can continue to perform their intended services.

Protection of underground structures from sabotage, theft, riot, unauthorized use, entry or other such action is very important to safeguard major investments in infrastructure.

Chapter 43
LEGAL CONSIDERATIONS

The practice of professional engineering including geoengineering continues to be affected by new legislation and relevant court decisions. In addition, the interest of governments in privatizing energy, water, and transportation projects, in North America and abroad, has led to new opportunities and challenges for engineers. These developments have generated keen interest in public/private partnerships and build-operate-transfer (BOT) projects, in which engineers play a major role. The expanding scope of risks and responsibilities undertaken on infrastructure projects increases the need for engineers to understand the importance of appropriate attention to legal issues.

Knowledge of the three basic forms of business organizations - sole proprietorships, partnerships, and corporations - is essential to the engineer's appreciation of legal rights and liabilities.

In a sole proprietorship, as the name suggests, an individual carries on business by and for himself or herself. The proprietor personally enjoys the profits of the enterprise and personally incurs any business losses of the enterprise.

The corporation is an entity unto itself, distinct from its shareholder owners. The corporation as an entity has been described as a fictitious person. The corporation itself owns its assets and incurs its own liabilities; it can sue or be sued in its own name. In fact, a shareholder of a corporation can contract with or sue that corporation.

Foreign markets are offering opportunities for engineers. The scope of the new world marketplace is vast. The project opportunities in China, India, Malaysia, Indonesia, Vietnam, and elsewhere in Southeast Asia, as well as South America and Eastern Europe are among the destinations of opportunity and challenge.

In both developing and developed countries lack of government capital to respond to pressing infrastructure needs has resulted in privatization opportunities for the private sector. These opportunities have arisen at a time when the North American development industries are in need of new markets. Accordingly, increasing numbers of engineers are responding.

New foreign markets generate new risks.

The choice of business organization plays an important role in other countries just as it does in North America. However, business organizations may be characterized differently in other jurisdictions, and tax and other issues will vary substantially from country to country. Accordingly, it is very important to obtain advice from an appropriately experienced lawyer in the foreign jurisdiction or to have the advice of a consultant familiar with the country.

Licensing Requirements

Compliance with licensing requirements and obtaining necessary permits and approvals in a foreign country are important considerations that can be time consuming. This is often the case when dealing with countries with a history of excessive bureaucratic procedures or inexperience in dealing with foreign investors on new types of project initiatives. The importance of local advice and relationships is a priority.

In the licensing process, a carefully chosen local advisor or local joint venture partner should be in a much better position to understand the realities and to deal effectively with bureaucracy than the foreign investor, consultant, or contractor.

Financial Risks

Many countries impose currency exchange controls or restrictions on the transfer of funds out of the country. Changes may also occur in import duties and result from local tax policy. These may all constitute significant risks related to foreign projects that need to be carefully investigated at the outset.

Contract Forms

A general observation of interest to North American engineers is that the contracting forms used on projects in many foreign jurisdictions are similar to their own contract forms. For example, the FIDIC contract form of the Federation Internationale des Ingenieurs Conseils, which are favoured for use on many projects financed by the World Bank, feature a third party consultant engineer authorized to make decisions similar to the approach contemplated by the Canadian CCDC contract forms. Countries such as China and Vietnam are, to some extent, basing contract approaches on Western forms and are taking advice on such forms from Commonwealth countries.

It is important for the engineer in business to understand the essential elements of a contract. For a contract to be binding and enforceable, five elements must be present:

a)   an offer made and accepted;
b)   mutual intent to enter into the contract;
c)   consideration;
d)   capacity to contract;
e)   lawful purpose.

Contracts consist of benefits to and obligations of the contracting parties. Agreements are generally arrived at by choice or through negotiation. The law will enforce the provisions of a valid contract. The law will not intervene to impose contract terms more favourable than those negotiated between the parties.

In certain circumstances the law may intervene to declare a contract void, voidable, or unenforceable, however, the engineer must be aware of one basic premise. If a bad business deal is negotiated, the courts will not impose more favourable terms.

The doctrine of fundamental breach may be applied to a contract that contains an exemption clause; essentially, it renders the exemption clause ineffective in the event of a fundamental breach of contract. An exemption clause is a provision whereby contracting parties may limit the extent, in whole or in part, of liability that arises as a result of breach contract.

A contract between a client and a professional engineer must include all of the essential contract elements.

A contract between a client and an engineer will not usually specify the measure of the standard of care in performance that is expected of the engineer. The contract will simply state that the engineer is to provide engineering services in connection with a particular project. The document may detail the scope of such services, but it will not necessarily specify the degree of care that is required of the engineer in carrying out those services. That degree of care will be an implied term in the contract. As pointed out earlier, an engineer is liable for incompetence, carelessness, or negligence that results in damages to the client. The engineer is responsible as a professional for not performing with an ordinary and reasonable degree of care and skill. The standard of performance expected of the engineer in contract is essentially the same as the standard expected in tort law, unless otherwise provided by a particular contract.

Most engineers are aware of the frequency of court actions against professionals. Professional engineers should carry appropriate and adequate professional liability insurance coverage.

Standard Form Engineering Agreements

Recommended forms of agreement for professional engineering services between client and engineer and between engineers and other consultants are available from state and provincial associations, and from other sources, such as a National Association of Consulting Engineers. The standard form engineering agreements currently available have been carefully developed. The forms set out very good basic contract formats. Some require more detail of completion than others with respect to project definition, the description of engineering services to be provided, and the engineer's fee for services rendered. Each contract should be tailored, so that it accurately records the particular agreement between the parties, a drafting process that is usually best left to lawyers or very experienced business managers.

In construction claims, North American courts have often dealt with cases involving issues relating to subsurface conditions. As difficult a dilemma as subsurface risks may present, legal cases confirm the strict approach the courts take to enforcing clear contractual obligations, even those undertaken in circumstances when the bidders may well have had far less than optimum conditions upon which to investigate, assess, and price the risks. However, where courts have decided in favour of the contractor claiming additional compensation on account of variations in subsurface ground conditions, a key ingredient has been substantiating that the project owner and/or engineer failed to disclose important information to the bidders.

The failure of an owner to disclose important information to bidders is now recognized as a basis for a contractor to claim additional compensation, subject of course to agreed contract terms.

In spite of the contract provisions relating to subsurface conditions, sometimes project owners contractually shift the subsurface risks to the contractor, to a very substantial degree. Owners who do so risk substantially inflated bids by contractors seeking to protect themselves from unknown subsurface conditions often referred to as a high price contingency to attempt to cover the risk.

Arbitration

Implementation of the ADR techniques, including partnering, mediation, dispute review boards, and arbitration makes a great deal of sense given the history of costly, time consuming, and disruptive construction litigation in North America and the West.

Professional Engineer's Seal

In most of the common law jurisdictions in North America, a professional engineer is required to stamp drawings and specifications with his or her seal. The seal is issued by the state or provincial association, and it indicates that the engineer is a registered Professional Engineer. It is extremely important that each engineer closely controls the use of that seal and ensures that it is only used where appropriate. Improper use of the engineer's seal can result in disciplinary proceedings and very substantial fines for individuals, corporations, and partnerships.

Penalties

The offence provisions of the statutes that regulate engineering impose varying penalties for contravening statutes and ordinances. The penalties are high for practicing professional engineering without a license, or for holding oneself out as engaging in the practice of engineering without being properly licensed.

Professional engineering is mostly a self governing profession. The disciplinary process provides sanctions that can be applied independently of any lawsuit and is therefore a very important basis upon which the state and provincial bodies may govern, in the interests of maintaining professional standards to ensure that the public interest is served and protected.

Certificates of Authorization

Professional engineering membership alone does not qualify engineers to offer to the public, or engage in the business of providing to the public, services that are within the practice of professional engineering. A certificate of authorization is also required. Applicants for certificates of authorization, including individual members, partnerships, and corporations, must meet prescribed requirements and qualifications pursuant to the Professional Engineers Act in the state or province.

The Code of Ethics

The Codes of Ethics under the Professional Engineers Act in a state or province and as endorsed by any National Council of Professional Engineers provide for appropriately high standards of duty, conduct, and integrity. Such high standards are very important from a technical perspective given the responsibility assumed by, and integral to, the undertakings of Professional Engineers as guardians of the public safety. It is important that Professional Engineers respect and implement their codes of ethics as professionals discharging their duties to the public, their employers, their clients, their colleagues, their profession, and themselves. Engineers have long fulfilled important leadership roles in society and enjoy a high degree of respect because of the achievements, abilities, and integrity engineers have demonstrated. Maintaining a strong sense of duty and fulfilling the expectations of the Code of Ethics is important to maintaining the strength and esteem of the professional engineering profession society.

Professional Engineers, as determined by education, training, licensing, and/or experience are held to a standard or ordinary skill in their rendering of service. The services of experts are sought because of their special skills. They have a duty to exercise the ordinary skill and competence of members of their profession and a failure to discharge that duty will subject them to the liability of negligence. Lawsuits against professionals are generally categorized under the term malpractice.

Over the years, earthworks and foundation claims have become considerably more sophisticated in terms of technical arguments, and they are prime issues of dispute resolution. Analyzing and evaluating the facts, in a straightforward and understandable way, and presenting an effective defense against these claims require special skills. The effectiveness of these efforts governs the success or failure of the respective parties.

Potential Plaintiffs

While most professionals, such as doctors, lawyers, or dentists can be sued only by their clients for their failure to properly perform and in rare cases, a third party, the engineer must protect himself and defend his professional conduct against claims by:

The Client
A subsequent purchaser
The general contractor
Subcontractor
The contractors' or subcontractors' employees
Governmental or quasi-governmental agencies
Members of the public
Other professional parties to the performance of the work, such as architects, structural engineers, landscape architects, etc.
Independent groups such as conservation societies and material or component suppliers

This large group of potential plaintiffs creates a special problem for engineers. For geotechnical engineers, the list of potential plaintiffs can be even longer.

Liability Suits

Most liability suits against an engineer start with the familiar sentence that the engineering firm has not met the standard of care. The key confusion as to the term standard of care is as follows:

Professional liability is, in fact, an objective standard imposed upon the professional and measured by a reasonably prudent practice for those engineers in similar activities and in the same geographic area. Thus, it is obvious that the standard changes from time to time and from place to place. What is acceptable practice today may not be acceptable tomorrow or may not be acceptable today in another community.

Frivolous Suits

One of the primary concerns of the design professionals in private practice is the lawsuits that have no merit or factual basis. Such unfounded lawsuits are commonly referred to as shotgun suits or frivolous suits, e.g., the plaintiff names anyone and everyone remotely connected with the occurrence that caused the injury.

Comparative Negligence

As a general rule, when an accident occurs on a construction site, everyone connected with the project is brought into the resulting suit. Under the contributory negligence rule, the plaintiff should not recover damages that they caused themselves by retaining the 50% cutoff for recovery.

In comparative negligence jurisdictions, the plaintiff's own negligence may not bar him from obtaining any recovery at all. The owner, however, is only able to recover a judgement for the percentage of the claim that corresponds to the degree of capability of the defendant for the injuries.

Joint and Several Liability

Another issue is that joint and several liability rule. This rule, holds that where several defendants, either in concurrence or independently, cause damage to the plaintiff, the plaintiff can recover entire damage awards from any one defendant, usually the one with the deep pockets such as sole insured.

Punitive Damages

Punitive damages in pain and suffering award have occurred in suits against engineers, such as in the case where the actual damage is small, but the amount of punitive damage is significantly higher than the repair cost.

Changes of Conditions

The change of conditions clause, or as it is now often called, the differing site conditions clause, is one of the most significant risk allocation clauses found in a construction contract. Without the clause, the contractor bears most of the financial risk associated with the encountering of onerous job site conditions unforeseeable at the time of bidding. With the clause, the owner accepts that risk.

On the other hand, for a highly competitive bidding contract, some contractors may intentionally lower the bid in order to get the contract. They expect to recover any loss through arguing about the site conditions. With the change of conditions awarded to them, they may gain financially.

There is no limit to the factual situations giving rise to a changed condition. Typical of the conditions that have been found to qualify as changed conditions that concern geotechnical engineers are:

a) The presence of rock or boulders in an excavation area where none or few were shown on the borehole log.

a) The encountering of rock or boulders in materially greater quantities or at different elevations than indicated in the borehole log available to bidders.

b) The encountering of groundwater at higher elevations or in quantities in excess of those indicated in the data furnished to the bidders.

d) The difficulty in adopting a drilled pier system as recommended.

e) The elevation of bedrock.

f) The necessity of using casings to complete the drilled piers.

Most engineering contracts are provided with clauses that stipulate the geotechnical report attached is for the use of the bidder as reference. The bidder should conduct his or her own investigation to verify the accuracy of the document. However, in a court of law, the owner cannot avoid the responsibility of the information provided to the bidders.

Expert Witnesses

In most legal cases, expert witnesses are required. This is essential when engineering practice is involved. The court must establish that a case exists and must prove that the defendant is the party responsible for the said damage and should be held accountable. It is obvious that most attorneys have little knowledge of engineering and must rely on experts to analyze the problem.

Geotechnical engineers may be asked to testify as expert witnesses in lawsuits, an activity that normally is voluntary and for which the engineer is duly compensated. This is in contrast to a subpoena, which is served by an officer of the court and requires that a witness will appear and testify at a specified time and date. During the trial, the judge as well as the jury must rely on experts to explain the problem in layman's language.

Clearly, in a complicated engineering project, it is a battle between experts. Unfortunately, most engineers have little or no experience in serving as expert witnesses. When they first appear in court or at deposition, they often find the experience far different from any they have ever known. Their prior experience in giving technical reports does not prepare them for the possible jolt their egos may undergo while appearing as expert witnesses.

Standard of Care

Before agreeing to serve, one must ask if there is an understanding of prevailing standard of care as it applied to the circumstances at issue. One must know what the average engineer might reasonably have done under similar conditions before an expert opinion can be formed about what, in fact, was done. One must think in terms of ordinary skill and care; if not, one is probably the wrong expert. The key questions a plaintiff's attorney will ask are as follows:

"Has the engineer in his/her work employed that degree of knowledge ordinarily possessed by members of that profession, and to perform faithfully and diligently any service undertaken as an engineer in the manner a reasonably careful engineer would do under the same or similar circumstance?"

This is a lengthy question. In fact the attorney is asking not what would have been done, but if what has been done was reasonable at that time and in those circumstances. If the answer is "no," the defendant is doomed. If the answer is "yes," the case will continue, and the defendant has a good chance. Remember, one may not agree with his or her conclusion and presentation, but he or she still has exercised the standard of care.

The best solution, whatever that might be, is not the issue. The issue is usual and customary care. One must differentiate between reasonable care and substandard performance. This puts the defendant in a favourable position. As an expert witness, it can put someone on the spot, because serving as an expert can be a challenging task. It is not to be taken without careful consideration.

It is therefore important to understand that a consultant gives opinion and advice but does not guarantee performance. It is up to the insurance company to guarantee performance. For average projects, the owners pay the consultants only a fraction of what they pay for the insurance.

Steps Leading to the Courtroom

a) Upon being approached by a client's attorney, the engineer can ask sufficient questions concerning the case to determine if it is within his or her area of expertise, and if the claims may have merit. The expert can indicate a willingness to serve without prejudice, and emphasize that he or she will call it as they see it. Most clients and attorneys will respect this position because they want the facts of the matter. Full and honest disclosures result in most cases being settled out of court.

b) The engineer will supply the attorneys with a curriculum vitae showing education and experience. Many federal courts require that the expert supply opposing counsel with a list of all cases, including attorneys' names, in which the witness has given testimony in the last five years. There sometimes is a rush to a particular expert to prevent his or her being retained by the other side.

c) At the request of the clients counsel and prior to giving testimony, the expert will review all available information including reports and legal depositions of others, and if requested prepare a written report analyzing and explaining the information. The attorney may ask that anything in writing carry a written notation, 'Privileged and Confidential Attorney-Client Document' because preliminary studies can lead to new areas of investigation, and as additional evidence is compiled opinion may change. As many of the same documents will be used by both sides there can be a duplication of materials.

d) Discussion with the client or client's counsel may lead to additional testing to try and help resolve the conflict. The client should be aware that test results can work either way.

e) The witness will discuss his or her findings with the attorney and client. A written report will include various documents, drawings, calculation sheets, maps, etc. used in rendering an opinion. When these are finalized they will become formal exhibits and given exhibit numbers. The report itself will become a court document.

f) In contrast to popularized versions of courtroom trials, attempts will be made to expose all relevant information for review by both sides prior to trial so that there will be no surprises. The most common method for accomplishing this is to require depositions under oath before attorneys representing all sides of a conflict. The deposition is requested and paid for by the opposing side(s). Opinions expressed in the deposition should be documented and based on the evidence and not on opinions or preferences of the client.

g) If matters are not settled, testify at trial.

Mediation and Arbitration

If an agreement is not reached, in order to save costs of a trial both sides may agree to hire an independent mediator or team of mediators, which will include both engineering and legal representatives, to try and make recommendations that are equally dissatisfying to both sides. The recommendations may or may not be accepted.

A more stringent approach is called binding arbitration. In this case a single independent arbitrator or a team of three or more, always an odd number to prevent a tie, will reach a decision that is binding. Arbitrators are nominated by each side, and at least one will be an independent investigator who will analyze the various arguments and may carry most of the weight of the decisions. The arbitration procedure may require a formal record and testimony under oath, or it may simply involve a review of the evidence and exhibits by the expert or panel of experts.

Testifying Under Oath

Of the various steps, the legal description can be the most challenging because attorneys will be probing for soft spots and inconsistencies. Prior to the deposition, the client's attorney will discuss the deposition protocol and give hints as to what the witness may expect. He will offer suggestions on how to give a good deposition, such as waiting until an entire question has been asked and not simply shaking one's head to answer a question. The discussion will focus on the issues, and the well informed attorney will know the answers to questions before they are asked.

A deposition is a formal question and answer procedure that begins with swearing in of the witness, who will be asked to state his or her name and affiliation. The opposing attorney(s) then ask questions of the expert, which is the direct examination. Attorneys on both sides will alternate questioning to seek additional clarifications.

An official record of every word spoken is typed in shorthand except for moments that are requested by an attorney to be off the record. Answers should be carefully considered based on the evidence, and never hurried. An answer can carry the qualifier, 'It is my opinion that…'

At trial the order of questioning is reversed, with the attorney for the client doing the direct examination and the opposing attorney the cross-examination.

In many ways a deposition and trial have elements of a game, so the witness should be relaxed and remember it is only a game. Each attorney is ethically bound to represent the best interests of his or her client, so while the evidence that is presented must be truthful, counsel may prefer that not all of the evidence be presented. In theory, with equally competent representations, the truth will come out and the right side will win.

Witness Demeanor

Expert witnesses have some special privileges, the most important being that they are not always required to give a simple yes or no answer but are free to explain the basis for their opinions. The expert witness should be a teacher, not an advocate or a combatant. Graphics often are used and are prepared ahead of time. During trial the judge can ask for clarifications.

Some trial attorneys resort to simple tricks, like asking the same question over and over again to test if the witness is consistent with the answers, or may be overly consistent by giving the same answer word for word, implying that it has been coached and rehearsed. A report may be read sentence by sentence and the witness asked if he agrees with each statement, even though it may be the omissions that are relevant. The expert inevitably will be asked if he or she is being paid for his or her testimony, which is intended to demonstrate if the witness is a hired gun. The appropriate answer is that one is being compensated for one's time, the same as the attorneys.

Evasive answers, bluffing, name dropping, inconsistencies, and technical jargon destroy witness credibility. Lying under oath is perjury, can impeach a witness, and carries stiff penalties. Opinions should be carefully phrased; for example, it would be unwise to slam water witching if the township's leading water witching person may be on the jury and it has nothing to do with the case. More accurate would be to say that one is not aware of any scientific validation for water witching so its reliability is open to questions. During trial the witness will be asked to explain any apparent deviation from his or her earlier reports and depositions.

Attorney smirking or ridicule can be expected, but only indicates that the opposing attorney is not comfortable with the way things are going.

The expert often is asked if in his or her opinion an investigation or report meets professional standards that were in place at the time that the report was written, which in many jurisdictions can form a basis for a decision.

Attitudes

At trial the expert who really is expert will speak slowly and clearly, use simple terms that are easily understood by lay persons, and explain technical terms as necessary. In an ideal world the experts would not take sides and would come to the same conclusions based on the same evidence, but this seldom happens. However, the disagreements generally are over interpretations and relevance of particular information or data. In rare cases experts on both sides will be asked to work together and issue a joint opinion as a basis for settlement.

Chapter 44

OUR EARTH'S FUTURE

Recognizing that our Earth is 4.6 billion years old, the last century of human civilization has seen fast and incredible advances in technology based on our science, engineering and business skills. Industrial revolutions have brought an abundance where consumerism versus sustainable development has preoccupied our societies. This can have a serious impact on our environment and the Earth.

Some scientists feel that the relationship between human civilization and the Earth is being affected by the population explosion, the technological revolution and a willingness to ignore the future consequences of our present actions.

Many governments of the industrialized countries have promoted reduce, reuse and recycle programs to help, and communities and individuals are responding.

Al Gore, in his Inconvenient Truth states,

> "In every corner of the globe – on land and in water, in melting ice and disappearing snow, during heat waves and droughts, in the eyes of hurricanes and in the tears of refugees - the world is withering, mounting and undeniable evidence that natures cycles are profoundly changing."

A planetary warning of global warming is demonstrated. A thin layer of Earth's atmosphere is being thickened by large quantities of human caused carbon dioxide and other greenhouse gases. As it thickens, it traps a lot of the infrared radiation that would otherwise escape our atmosphere and continue out to the universe. As a result, the temperature of the Earth's atmosphere and oceans is becoming warmer. The insurance industry has seen very significant increases in claim payments to victims of extreme weather including hurricanes, floods, drought, tornadoes, wildfires and natural disasters associated with global warming.

Global warming may seem gradual in the context of a single lifetime, but in the context of the Earth's natural history, it is actually happening with speed.

Soil moisture evaporation or soil dehydration increases with higher temperatures is causing less productive agriculture and more building settlement and cracking, plus other ill effects.

It is important to work on alternatives to fossil fuels and to begin to break our dependence on foreign oil by using national energy sources such as biomass, solar, wind, hydro and geothermal. If global warming continues whereby Greenland melted or broke up and slipped into the sea, then sea levels would increase by about 6 m (20 ft) and the maps of our Earth will have to be redrawn. For example places like South Florida and the Netherlands are predicted to be submerged.

Currently we have the ability to transform the surface of the Earth. Every human activity is undertaken with ever more powerful tools which can bring unanticipated consequences such as with deforestation, open pit mining, industrialization, urban sprawl and war munitions.

In studies at Princeton University it was found that humanity possesses the fundamental scientific, technical and industrial knowledge to solve the carbon and climate problems by the following approaches:

- Reduction from more efficient use of electricity in heating and cooling systems, lighting, appliances and electronic equipment

- Reduction from end-use efficiency, meaning that we design buildings and businesses to use far less energy than they currently do

- Reduction from increased vehicle efficiency by manufacturing cars that run on less gas and putting more hybrid and fuel cell cars on the roads

- Reduction from making other changes in transport efficiency, such as designing cities and towns to have better mass transit systems and building heavy trucks that have greater fuel efficiency

- Reduction from increased reliance on renewable energy technologies that already exist, such as wind and biofuels

- Reduction from the capture and storage of excess carbon from power plants and industrial activities

As Al Gore states,

"the truth about the climate crisis is an inconvenient one that means we are going to have to change the way we live our lives."

Similarly Professor David Suzuki is interested in the interface between science and society. He feels that throughout the history of Earth our civilization is deeply dependent on the natural world. As human numbers have exploded, more of the Earth's population have spent their entire lives in a modest one hundred year period of unprecedented growth and change, and naively consider this as a normal condition of humanity. It is expected, but it is not sustainable development.

In considering the important balance between our economy and our planet, Earth, an important consideration is that the good health of our environment should precede business and commerce, otherwise there can be no future economy. At this time, needed environmental improvements will eventually lead to profitability and prosperity.

Knowing that geoengineering sciences and applications play one of the most significant roles in the taking and return of Earth's natural resources is it a reasonable mission in earth engineering and sustainable development to consider such a pledge to planet Earth as follows:

The Declaration of Interdependence (Suzuki 2010)

This We Know

We are the Earth, through the plants and animals that nourish us.
We are the rains and the oceans that flow through our veins.
We are the breath of the forests of the land, and the plants of the sea.
We are human animals, related to all other life as descendants of the firstborn cell.
We share with these kin a common history, written in our genes.
We share a common present, filled with uncertainty.
And we share a common future, as yet untold.
We humans are but one of thirty millions species weaving the thin layer of life enveloping the world.
The stability of communities of living things depends upon this diversity.
Linked in that web, we are interconnected - using, cleansing, sharing, and replenishing
the fundamental elements of life.
Our home, planet Earth, is finite; all life shares its resources and the energy from the Sun,
and therefore has limits to growth.
For the first time, we have touched those limits.
When we compromise the air, the water, the soil, and the variety of life, we steal from the
endless future to serve the fleeting present.

This We Believe

Humans have become so numerous and our tools so powerful that we have driven fellow creatures to extinction, dammed the great rivers, torn down ancient forests, poisoned the Earth, rain and wind, and ripped holes in the sky.
Our science has brought pain as well as joy; our comfort is paid for by the suffering of millions.
We are learning from our mistakes, we are mourning our vanished kin, and we now build a new politics of hope.
We respect and uphold the absolute need for clean air, water, and soil.
We see that economic activities that benefit the few while shrinking the inheritance of the many are wrong.
And since environmental degradation erodes biological capital forever, full ecological and social cost must enter all equations of development.

We are one brief generation in the long march of time; the future is not ours to erase.
So where knowledge is limited, we will remember all those who will walk after us, and err on the side of caution.

This We Resolve

All this that we know and believe must now become the foundation of the way we live.
At this turning point in our relationship with Earth, we work for an evolution: from dominance to partnership; from fragmentation to connection; from insecurity to interdependence.

The notion that the Earth is just one of a number of planets orbiting the Sun is both fundamental and unsettling. It removes the Earth, and more importantly, the human race from the centre of our universe and vast galaxies. Cosmology explains the origin and nature of our universe. Geology provides an important understanding of the character of our dynamic planet. Earth engineering provides tremendous geoscientific knowledge and ingenious geoengineering solutions for mankind and life to progress. In another 50 to 250 million years of geological time continental drift will vastly change the maps of the Earth, likely to include no polar ice caps and the Mediterranean Sea closing.

In the way that ants have loyalty to their colony, we are the important caretakers of our home planet. All loyalty to sustainable earth engineering is based on two elements – the hope of protection and the hope of enhancement. At the most down to earth level of self interest, we realize that the totally continuous and interdependent systems of air, land and water maintain and evolve the life on our planet. Advanced geosciences explore how Earth depends for its survival on the balance and health of our ecosystems. A wider based loyalty by nations to sustainable earth engineered developments is needed. A unified recognition of our geoenvironmental interdependence gives us a sense of community, of belonging and living together, without which no human society can be built up, survive and prosper.

Our future is intimately connected with that of our planet. Earth is uniquely beautiful and fascinating. The human race is crafted to life on Earth as a result of more than 1.5 billion years of evolution here by us and our ancestors. This book provides several essential perspectives, principals and practices of earth engineering for sustainable development.

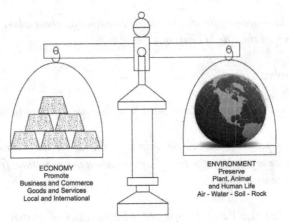

ECONOMY
Promote
Business and Commerce
Goods and Services
Local and International

ENVIRONMENT
Preserve
Plant, Animal
and Human Life
Air - Water - Soil - Rock

44.1 Earth in Balance

Partial List of References

American Society for Testing and Materials (1981). Permeability and Groundwater Contaminant Transport. Special Technical Publication (STP) 746. ASTM, Philadelphia

American Society for Testing and Materials (Published Annually). Section 4, (for 1984) Construction: special reference to Vol. 04.01 Cement; Lime; Gypsum. Vol. 04.02 Concrete and Mineral aggregates. Vol. 04.03 Road and Pavement Materials. Vol. 04.08 Natural Building Stones, Soil and Rock. ASTM, Philadelphia.

ASCE, Lateral Stresses in the Ground,

Specialty Conference, Cornell University, 1970

ASCE, Design and Performance of Earth Retaining Structures Geotechnical Conference SP.No. 25, Cornell University, 1990

Andersland, O. B. and Ladanyi, B., (1994) "An Introduction to Frozen Ground Engineering", Chapman & Hall, New York

Baird, Colin, "Environmental Chemistry," 2nd Edition, W. H. Freeman, 1999, New York

Bell, F. G. (1993) Engineering Geology, Blackwell Scientific Publications, Canada

Bles, J. L., and Feuga, B. "The Fracture of Rocks," Elsevier, 1986

Blyth, F.G.H. and DeFreitas, M.H. (1984) "A Geology for Engineers," 7th Edition, Elsevier, New York

Brady, B.H., and Brown, E.T. "Rock Mechanics for Underground Mining," George Allen and Unwin, 1985

British Standards Institution (19720. Code of Practice for Foundations, CP2004: 1972. British Standards Institution, London.

Brown, E.T., editor, "Rock Characterization, Testing and Monitoring," ISRM Suggested Methods, Pergamon Press, 1981

Corns, C. F. (1974), "Inspection Guidelines, General Aspects. Safety of Small Dams," Proceeding of Engineering Foundation Conference

Fairbridge, Rhodes, W., "The Encyclopedia of Geochemistry and Environmental Sciences," Van Nostrand Reinhold, 1972

Farmer, I., "Engineering Behaviour of Rocks" 2nd Edition, Chapman and Hall, 1983

Franklin, J. A., and Maurice B. Dusseault, "Rock Engineering," McGraw Hill, 1989

GeoLogan 97 Conference (1997) "Grouting: Compaction, Remediation and Testing," Geotechnical Special Publication No. 66. GeoInstitute, ASCE, New York

Gillot, J. E. (1987) "Clay in Engineering Geology," Developments in Geotechnical Engineering, Vol. 41, Elsevier, New York

Goodman, R.E., "Introduction to Rock Mechanics" John Wiley & Sons, 1980

Goodman, R.E., "Methods of Geological Engineering," West Publishing, 1976

Hartman, Howard L., "Introductory Mining Engineering," John Wiley & Sons, Inc., 1987

Houghton, J. T., (1997), "Global Warming," 2nd Edition, Cambridge University Press, Cambridge, U.K.

Huang, Y.H., (2004), "Pavement Analysis and Design," 2nd Edition Prentice Hall, New Jersey

Fluet, J. E., "Geosynthetics for Soil Improvement: General Report and Keynote Address," Geosynthetics for Soil Improvement, R. D. Holtz editor, GSP No. 18 ASOE, New York, 1988

Hoek, E., and J.W. Bray, "Rock Slope Engineering," 2nd Ed. The Institution of Mining and Metallurgy, London, 1977

Huang, Yang H., "Stability Analysis of Earth Slopes," Van Nostrand Reinhold Co. Inc., 1983.

Hudson, J.A., "Rock Mechanics Principles in Engineering Practice," London,1989 Construction Industry Research and Information Association, Butterworths

James, David E., "The Encyclopedia of Solid Earth Geophysics," Van Nostrand Reinhold, 1989

Jumikis, A.R., "Rock Mechanics," 2nd Ed., Trans Tech, Karl, 1983

Koerner, R., "Designing with Geosynthetics," 4th Edition, Prentice Hall, Englewood Cliffs, N.J. 1998.

Legett, R. F. and Hatheway, A. W. " Geology and Engineering," 3rd Edition, McGraw Hill, 1988

Legget, R. F. and Karrow P. F. (1983) "Handbook of Geology in Civil Engineering," McGraw-Hill Book Co., New York London

Mandl, G., "Rock Joints, The Mechanical Genesis," Springer, Austria, 2005

Mathewson, C. C. (1981) "Engineering Geology," Charles Merril Publishers, Columbus, Ohio

Middlebrooks, T.A. (1953) "Earth Dam Practice in the United States" Transactions, ASCE, Centennial Volume, pp. 697

Nelson, P.P., and S.E. Laubach, editors, "Rock Mechanics, Models and Measurements, Challenges from Industry," Balkema, Rotterdam, 1994

Papagiannakis, A.T., "Pavement Design and Materials," John Wiley & Sons, Inc., 2008, Hoboken, New Jersey

Pinder, George Francis, "Groundwater Modelling Using Geographical Information Systems," John Wiley & Sons, 1942, New York

R. W. Day, "Geotechnical Engineers Portable Handbook," Appendix E

Rahn, P. H. (1986) "Engineering Geology, An Environmental Approach," Elsevier, New York

Rollings, M.P, and Rollings, R.S., "Geotechnical Materials in Construction,"McGraw-Hill, 1996

Sharma, P.V., "Environmental and Engineering Geophysics," Cambridge University Press, UK, 1997

Sherad, J. L. et al (1963), "Earth and Earth Rock Dams" John Wiley & Sons, New York, 725 pp.

Sowers, G. F. (1974), "Dam Safety Legislation: A Solution or a Problem" Safety of Small Dams, Proceeding of Engineering Foundation Conference

Tomlinson, M. J. (1980). "Foundation Design and Construction," 4th Edition. Pitman Publishing Ltd., London

United States Bureau Reclamation (1977). Groundwater Manual: A guide for the investigation, development and management of groundwater resources. Denver, Colorado

Waltham, A.C., " Ground Subsidence," Blackie UK, Chapman and Hall USA, 1989

Walton, W. C. (1970), "Groundwater Resource Evaluation", McGraw-Hill, New York

Zhang, L., "Engineering Properties of Rocks," Elsevier, 2005